IN PRAISE OF
MOSAIC OF MINDS

I imagine a future society where we live together with highly intelligent AI. If we could determine our future, what would it look like? Would our deeply held values and ethics persist? What is the ideal future, and what will be the foundation for it? Even after spending considerable time developing cutting-edge AI in Silicon Valley, I still don't have the answers to these questions. *Mosaic of Minds* offers us a guiding light toward a more humane future amidst the uncertainty. As you turn the pages, you may encounter questions you've never considered before—questions we must ask if we want to navigate the ambiguity of our future. We are fortunate to encounter a book brimming with warm visions and profound insights into the future during these turbulent times.

—Cho Sung-Jung
Engineer, Meta

Former Engineer, Google

Love? What's your price?" This line from the drama series *Autumn in My Heart* still lingers with me after twenty or so years. It feels strange and forced to try and convert love into money, a human value into something specific and quantifiable. Quantifying human values has long remained a challenging and complex task. However, with the advent of AI technology, this interesting yet perplexing topic has come to the forefront. While we still lack a consensus on what constitutes being human, we seek to quantify our humanity through AI. For instance, some students who condemned the unethical nature of Facebook had no ethical qualms when working on projects for social media companies that involved exploiting user profiles. As such, our ethical compass varies by context, so we must contemplate on how we can approach and unify our ethical views when building a humane future with AI technology. Through discussions with scholars of various fields, *Mosaic of Minds* lays out the insights of those who have already thought of the above issues. Interestingly enough, despite their diverse backgrounds and fields of study, their thoughts and insights seem to align. This book is a map that can help us navigate through the complicated yet essential problem that needs to be solved.

—Chun Byung-Gon

Professor, Computer Science and Engineering Department, Seoul National University

CEO, FriendliAI

Upon first seeing the title, I thought it could have been *New World and Ethics of New Technology*, but it used "humanity" instead. Why? Perhaps the author envisioned a future where a person, enjoying the fruits of new technology (AI), or maybe a child in pain because of it, would be affected. If one realizes how something can impact the life of a child decades later, their attitude toward that technology can never be the same. Rules, dangers, dilemmas, and overcoming prejudice in a new world ... An engineer who sought only the most optimal algorithm joins hands with a philosopher to draw out their map, thinking of the world they will leave behind. From then on, the engineer's goal is not stock options, but an algorithm whose beginning may be forgotten, but could bring happiness closer to all. I want the readers of this book to envision the same: a post-apocalyptic world. There, on a desolate wasteland, walks a nomad who knows not what the new world's fair principle is. They stumble upon a grand settlement in the middle of the desert, where a group of philosophers, sociologists, computer scientists, entrepreneurs, and politicians who understand both technology and philosophy research the ethics of the new world on equal grounds, welcoming them warmly. Imagine sitting at the end of their table with a warm meal, in the seemingly last bastion of mankind, engaging in discussion with them.

—Kim Jong-Woo
Production Director, MBC, Meeting You

The recent advancements in AI are truly remarkable. As a professor teaching AI to students and as a researcher at the forefront of AI research and development, I take great pride and feel a sense of anticipation in these achievements. However, the rapid development and incredible results of AI also instill a sense of fear and responsibility in me. Amidst the glamour, this book encourages us to consider the impact AI will have on our society. As we enter an era that requires careful thought and action regarding AI ethics and social responsibilities, this book offers valuable insights and guidance. I wholeheartedly agree with the argument that discussions involving experts in humanities and sociology should take place during AI research and development to establish a common ethical foundation. The numerous efforts and initiatives by institutions like Stanford, Harvard, and MIT can serve as significant examples for us to learn from and build upon.

—Lee Seong Whan
Professor, Department of Artificial Intelligence, Korea University

As AI becomes an integral part of our lives, the ethics surrounding AI have gained global attention. The EU announced their AI regulation bill in 2021, and the South Korean government unveiled their ethics guidelines, with companies like Naver, Kakao, and NCSOFT also announcing their own AI ethics regulations. I agree with the idea suggested in this book that the boundaries between fields of study must be dissolved to create AI that benefits humanity. As we approach a society where we interact with AI, its infrastructure must be built upon sociology, philosophy, and ethics, requiring ongoing discussions among developers, users, and governments, and a general consensus within our societies. This book proposes a new framework for humans and AI to coexist, which will become an intellectual resource for navigating the complexities and uncertainties of our era. Most importantly, it provides insights not only for businesspeople but also for those in education, humanities, sociology, government policy makers, and the younger generation who will shape the future.

—Ha Jung-Woo
Head of AI Innovation, NAVER Cloud

"Life is already full of artificial intelligence." This statement is not an advertisement for an AI company; it reflects our current reality. We drive through routes guided by AI, purchase shirts recommended by AI, and listen to music selected by AI. AI can detect flaws and manage production lines in a canned fish factory. However, AI may still present a high mental barrier for some individuals. This uneasiness is likely to be temporary as we grow accustomed to AI. Our parents' generation found English challenging, but we now use words like "momentum" and "tier" in our everyday conversations. AI-related knowledge will follow a similar path. Naturally, there are ways to adapt more quickly to the age of AI, and this book serves as an excellent guide. The book poses broad yet profound questions such as "what should universities teach to cultivate AI engineers with a strong ethical foundation?" and "what would Descartes think of current AI if he could witness it?" Most importantly, it helps us understand how

AI, which originated from computer science, has merged with fields like economics, education, and government. I recommend this book to those who wish to see the bigger picture of change and overcome the information overload of AI news. It is particularly recommended for my fellow journalists, who are in great need of AI literacy.

—Kim Tae Gyun
Team Lead, Yonhap News Contents Incubation

AI, in its narrow sense, may refer to algorithms or software like AlphaGo, but from a broader perspective, it serves as the foundation of our future society. As a result, AI is being redefined based on each individual's understanding and interpretation. This book acts as a guide to showcase various aspects of AI and facilitate comprehension. Since we expect AI to evolve by imitating our thinking abilities, it is considerably more challenging to predict the future it will bring compared to the changes introduced by steam engines, computers, or cars. While many movies and books depict dystopian futures revolving around AI, we cannot be entirely negative or optimistic about a future with AI. However, it is certain that we should approach AI with caution and aim for coexistence. Bearing this in mind, the title of this book, *Mosaic of Minds*, represents the authors' goal, and the book contains insights on questions we must address to achieve a human-centered AI. This is a must-read for everyone who will need to live through an uncertain future.

—Lee Yoon Geun
Research Lab Director, ETRI AI

As someone pondering the direction of AI research in our country, the question I am most frequently asked concerns the impact of artificial intelligence on our future lifestyle. This includes its effects on jobs, potential AI errors, malicious intentions of developers, and so on. The increasing concerns surrounding these issues underscore the growing need for AI regulations. When faced with these questions, I used to argue against the immediate need for control over AI technology development and suggested that discussions could take place over time. However, as I read through this book, I began to realize that there are many actions we can take right now. We can educate developers about their influence and potential biases during the planning and development of AI technology, instead of addressing problems after completion. Furthermore, we can engage in discussions with experts in ethics and philosophy, even though these topics may be difficult for developers to understand. This book explores our future coexistence with AI through conversations with experts from various fields, making it inspiring not only for those familiar with AI but also for those who study human lives and thoughts. Moreover, it provides an opportunity to engage in constructive discussions on the role of AI in shaping our humane future.

—Lee Hyunkyu

Information & Communication Technology Planning & Evaluation (IITP), Ministry of Science and ICT

Project Manager, Artificial Intelligence

Mosaic of Minds

Mosaic of Minds

NAVIGATING

THE COEXISTENCE OF

ARTIFICIAL INTELLIGENCE

AND HUMANITY

SONGYEE YOON

 | Books

Published by Advantage Books, Charleston, South Carolina.
An imprint of Advantage Media.

ADVANTAGE is a registered trademark, and the Advantage colophon is a trademark of Advantage Media Group, Inc.

Printed in the United States of America.

10 9 8 7 6 5 4 3 2 1

ISBN: 979-8-89188-027-6 (Paperback)
ISBN: 979-8-89188-028-3 (eBook)

Library of Congress Control Number: 2024901867

Cover design by Matthew Morse.
Layout design by Ruthie Wood.

This publication is designed to provide accurate and authoritative information in regard to the subject matter covered. It is sold with the understanding that the publisher is not engaged in rendering legal, accounting, or other professional services. If legal advice or other expert assistance is required, the services of a competent professional person should be sought.

Advantage Books is an imprint of Advantage Media Group. Advantage Media helps busy entrepreneurs, CEOs, and leaders write and publish a book to grow their business and become the authority in their field. Advantage authors comprise an exclusive community of industry professionals, idea-makers, and thought leaders. For more information go to **advantagemedia.com**.

To the pursuit of harmony and the ones who work to make it possible.

This book is based on the AI Framework series, in which NCSOFT engages with scholars from around the world to explore their views on AI technology and propose new perspectives.

CONTENTS

PROLOGUE

Humanity's Assignment, for the Most Humane Future

"What if we make the gender ratio of hero characters equal?"

While I cannot definitively state that all games do this, most game stories revolve around male characters, with the majority of core characters being male heroes. While this may seem gender biased from our current viewpoints, it is understandable considering the history of the game industry's development. It is undeniable that the majority of people in the industry, such as game designers, developers, and players, were predominantly men. However, as the number of female gamers continues to rise, we are confronted with many gender bias issues not only in games but also in various aspects of our society.

To draw attention to overlooked issues and stimulate discussion, I posed the aforementioned question to developers. However, the response I received was unexpected. Instead of acknowledging the need for change, their collective answer was, "Why do we have to?"

I was vaguely hoping for a positive reaction in the line of, "We were not fully aware of something so important," or "We must have missed it so we'll change it right away." So when they asked "Why?" I was flustered.

"Why is it better to balance character genders?"

"Why is it better to have balance?"

I came to the realization that I, too, had simply accepted things as they were without questioning why. I had not thought deeply about the answer or the reasoning behind it. This was the moment when I began to contemplate what is right or wrong, to continuously inquire about what truly constitutes right and wrong, and how to arrive at the answer.

AI is not only advancing faster than any technology we have seen, but it has also permeated our lives more naturally than expected. Just looking at the past couple years, the effect of AI technology has become incomparable, while its range has now reached beyond businesses and into politics, economics, and culture. AI is no longer just a field in computer science. It is a technology we face every day and a core part of our society connected to various aspects.

An AI that forms its perspectives and judgments of the world based on humanity's accumulated data could either perpetuate prejudice and bad habits stemming from our limited understanding or distorted viewpoints or, alternatively, build upon the wise choices that have sustained our civilization thus far to create a better world. As AI's influence over our society grows, concerns have mounted about a future where humans become irrelevant to machines. Nevertheless, periods of chaos and uncertainty make clear what we can do.

When confronted with an uncertain future, humanity has always clung to a delicate destiny, stumbling and wondering where to take the next step. Then, it firmly plants its foot on solid ground, extricates the other foot from the mud, and follows the path it has chosen, utilizing the most humanly feasible methods available.

To shape a humane future, we must pose questions, express concerns, and contemplate further to learn from our mistakes. We must tirelessly engage in debates about AI and with AI to discover a way forward. Every action taken by AI will reverberate throughout society; thus, it is critical that we gather diverse insights and discuss the optimal methods and directions we should pursue. Every viewpoint is valuable and necessary.

The current important and urgent tasks include continuously asking each other what could be better or what the right choice is and gathering thoughts on agendas that do not yet have social agreement. These tasks are not just academic research topics for AI experts, which can often seem disconnected from our daily lives, but are also rights for all of us who live alongside AI technology.

It is also our responsibility to shape our future society so that it heads in a more beneficial direction. If we don't want facial recognition technology, which is used to open your phone, to be used on surveillance drones peering through our windows, or if we don't want chats we had with our smart devices to be stored and used behind our backs by companies benefiting from them, or if we don't want an image we uploaded for fun to be edited and used against minorities to violate human rights, then we need more people to give detailed attention and constantly ask questions about living in a society that coexists with AI. While we must be wary of issues of fairness, freedom, and trust surrounding AI development, we must also be willing to discuss what is the better choice and more humane decision.

As part of my efforts to emphasize the importance of these tasks and encourage participation, I have advocated for the consideration of AI's impact and ethical issues from a business perspective. In addition, I have actively promoted governmental and nongovernmental discussions aimed at developing effective countermeasures. While it is necessary to persuade today's leaders to shift their focus from a quantitative and short-term perspective to a more sustainable and future-oriented approach, it is equally critical for young students who will become tomorrow's leaders to cultivate a habit of examining their lives based on philosophy and morality. They should be able to engage in discussions with their peers about whether their ideas could perpetuate past mistakes or have negative consequences for society. That's why I became affiliated with and supported the Embedded EthiCS program. During the process, I had many conversations and debates with renowned scholars, which were posted online last year. This year, I refined the content and published it as a book.

Of course, this book does not contain the secret method for dealing with all AI issues. Dealing with the current philosophical and ethical viewpoints surrounding AI is an ongoing flow of questions without answers. Discussions often become tangled with hard, complicated questions and multiple perspectives, which can be more confusing than informative. Nevertheless, I invite you all to join and be a part of this process. I hope my lengthy trail of thoughts and questions will pique the interest and curiosity of many more people. This is because I believe that we all desire a shared path toward a beneficial relationship between humanity and AI and that together we can find a solution that will help us create the most humane future possible.

We live in what many are calling the "post-COVID era," the new normal. The pandemic has swept away the framework we once believed in, leaving us in a time that is more chaotic and unpredictable than ever before. It is now time for us to fight back against uncertainty and disarray and find a common ground in our society by sharing our individual insights. There will be a variety of opinions, and we may not agree on a single approach from the outset. That's why we should not be complacent when it comes to discussing and debating. In fact, it might be better to embrace the mistrust, misunderstandings, and differences we have and then actively discuss what our inseparable AI partner should learn from the good values we have maintained.

I hope this book will serve as a catalyst to initiate debates and the sharing of opinions on establishing a social agreement regarding our fundamental values. I invite you to embark on a journey of discussions and inquiry together. Through our collective opinions, we can create a wise road map that can mitigate the risks brought on by AI and other emerging technologies and lead us toward a future that is beneficial to human society.

—Songyee Yoon

CHAPTER 1
The Future of the New Humanity

AI: Not Replacing but Augmenting Humanity

AI (Ethics) Framework × Fei-Fei Li

Dr. Fei-Fei Li is a computer science professor at Stanford University and co-director of the Stanford HAI (Human-Centered AI Institute). Previously, she served as vice president of Google and chief scientist of AI/ML for Google Cloud (2017–2018). She was recognized as one of Forbes' "8 Leading Women in AI" in 2020 and is a distinguished scholar in the AI field.[1]

Her current research focuses on computer vision, and she designed the world's largest image database for deep learning called ImageNet. In addition, she leads an initiative to implement AI into healthcare to address various challenges. Dr. Fei-Fei Li is also a co-founder of AI4ALL, a nonprofit organization that aims to promote diversity and inclusion in AI education, emphasizing STEM and early education for the development of AI education.

Focusing Our Viewpoints in the Face of Chaos and Fear

The AI singularity, predicted by futurists, in which AI surpasses human abilities, is approaching faster than expected. The COVID-19 pandemic, however, became an opportunity for development, resulting in the introduction of endless AI-related keywords, such as digital new deal, metaverse, self-driving vehicles, blockchain, cryptocurrency, and more into our society.

Undoubtedly, the rapid growth of AI technology has gifted human lives with greater convenience and productivity. However, with its increasing contribution to society, fears of AI overwhelming and replacing humans are also on the rise, with ethical debates surrounding AI coming to the forefront. The decision of whether AI is beneficial to humanity or whether its rapid development should continue remains inconclusive. Therefore, many computer science experts who pioneered the creation of AI have called upon people from various fields and societies to join the discussion. AI is no longer just a field of science but is now a technology that affects the entire world we live in today.

As we struggle through the chaos and uncertainty of our times, we are beginning to find a common ground. The development of AI technology presents both opportunities and challenges for humanity. If left unchecked, it could become a threat, but if we steer it toward benefiting humanity, it could become a tool for further advancement. It is widely agreed that all those involved in AI-related fields must come together and cooperate to ensure we stay on the right path.

Stanford University, a leader in computer science and AI, recognizes the importance of this consensus. In 2019, they established the Stanford Institute for Human-Centered Artificial Intelligence

(HAI) to facilitate cooperative discussions on developing AI that prioritizes human benefit. In this chapter, we speak with Dr. Fei-Fei Li, co-director of HAI, about the fundamental viewpoints and frameworks required for living in the age of AI.

I encourage you to align your perspectives with HAI, which has worked on various debates and collaborations to guide the tremendous power of AI toward advancing human society in a human-centered way. I hope you gain insight from how they have synthesized the diverse ethical debates surrounding AI into a common perspective.

The Need for Human-Centered AI

Yoon: Hello, Dr. Fei-Fei Li. Thank you so much for your time. The twenty-first century is rapidly becoming the age of AI, with this technology becoming the core and leading technology of the world. However, the innovative development of AI technology has resulted in its influence and proliferation being much faster than predicted, leading to considerable confusion. We have individuals who are barely aware that AI stands for artificial intelligence, while others drink coffee brewed by a robot barista and talk naturally with their AI assistant on their smartphone.

As such, awareness and preference over AI technology varies greatly between individuals, and we still don't have a definitive answer on whether AI will be beneficial or not. Society is thus in a melting pot of vague optimism and fear. It is intriguing that HAI took action in suggesting basic viewpoints and a framework for how we should handle AI technology, with a clear goal of creating human-centered AI.

As Stanford has been a leader in the field of AI since the beginning, their accumulated experience likely helped drive the initiative. Knowing that you, a professor of computer sciences at Stanford and the first co-director of HAI, have firsthand experience on its establishment makes our meeting more meaningful and eager. Out of all the numerous questions surrounding AI, understanding this strange technology, and determining our stance on it should be a top priority. I am thrilled to discuss these topics with you, Dr. Li.

Li: I am also happy to talk with you on these topics, Songyee. Thanks to advisors like yourself and many of our colleagues, HAI was able to grow into what it is today. I'd like to extend my gratitude for your support and participation of HAI's development. Reflecting on the past brings me back to when I first started in the computer science department tucked away in a corner of a large building. My colleagues were either computer science students or part of the AI lab. However, nowadays, AI is rapidly advancing and is driving significant societal change. We must acknowledge that AI is no longer solely a part of science and technology; rather, it is a multidisciplinary field that encompasses a diverse range of studies.

Yoon: That's correct. I, too, used to sit at the corner of the computer science department because AI has interdisciplinary characteristics that require the integration of many fields of study. In the past, computer science engineers did not view AI as a conventional field of engineering. Many misunderstood and suggested that we should be in psychology rather than computer science. We've come a long way since then. The number of computer science graduates in North America, including AI and robotics, bioinformatics, and big data analysis, has more than tripled over the last decade [see Table 1]. Moreover, the majority of recent computer science-related doctorates are AI/ML majors [see Table 2].

While all fields related to computer science and engineering have advanced greatly, AI has been unparalleled. I'm sure that it has been meaningful for you to not only witness the rapid growth of AI but also to lead the field. Your experiences have likely provided the foundation for your core insights as an AI expert, and your vision for the future of AI has been integrated into the mission and activities of Stanford's HAI. HAI is currently establishing a forum for discussions and experimentation to develop human-centered AI. The goal is to create changes in the real world by promoting collaboration among various fields in applying AI technology to public domains. I would like to ask how Stanford recognized the importance of establishing HAI, as well as what drove its actual creation. Additionally, could you provide an overview of the institute's history since its inception and its current focus in terms of goals and activities?

Li: The idea for what we now call HAI started about three to four years ago when AI technology began to emerge as a powerful force of change. Google DeepMind's AlphaGo defeated 9-dan Go player Lee Sedol, and self-driving cars were no longer just science fiction. At the same time, we became aware that AI-based facial recognition technology was biased, raising concerns about the accuracy and fairness of AI technologies. Privacy infringements were also taken very seriously, drawing attention to the importance of ethics in AI as its influence and technology continued to advance.

NUMBER of NEW CS UNDERGRADUATE GRADUATES at DOCTORAL INSTITUTIONS in NORTH AMERICA, 2010-20
Source: CRA Taulbee Survey, 2021 | Chart: 2022 AI Index Report

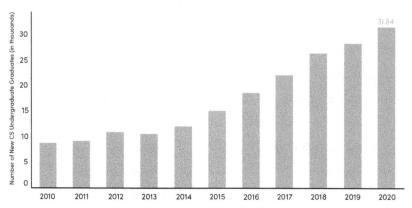

Daniel Zhang, Nestor Maslej, Erik Brynjolfsson, John Etchemendy, Terah Lyons, James Manyika, Helen Ngo, Juan Carlos Niebles, Michael Sellitto, Ellie Sakhaee, Yoav Shoham, Jack Clark, and Raymond Perrault, "The AI index 2022 annual report," AI Index Steering Committee, Stanford Institute for Human-Centered AI, Stanford University, March 2022.

As members of Stanford University, we realized that it was the right time to discuss our roles, opportunities, and responsibilities in this new age where AI technology was transforming society. It's worth noting that Stanford is home to one of the two oldest AI labs in the US, SAIL (Stanford Artificial Intelligence Laboratory), which was established in 1962. Professor John McCarthy, one of the lab's founding members, coined the term "artificial intelligence."[2] Since then, many members, including Turing Award winners,[3] self-driving car researchers, and deep learning pioneers, have been instrumental in advancing AI technology.

MAJORS OF COMPUTER SCIENCE DOCTORATE RECEIVERS OF 2020

NEW CS PHDS (% of TOTAL) in the UNITED STATES by SPECIALTY, 2020
Source: CRA Taulbee Survey, 2021 | Chart: 2022 AI Index Report

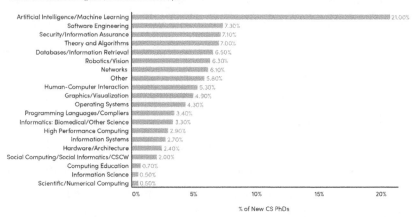

Stuart Zweben and Betsy Bizot, "2021 Taulbee Survey CS Enrollment Grows at All Degree Levels, with Increased Gender Diversity," accessed October 9, 2023, https:// cra.org/resources/taulbee-survey/.

My colleagues and I at Stanford took on the responsibility to plan for a new era where AI technology can provide a positive impact on human lives. We reminded ourselves that we must fulfill our duty to humanity as we continue to advance cutting-edge technology. This is why we strongly believe in developing and utilizing technology to improve the lives of everyone. Through participation and discussions among scholars at Stanford, HAI was born with the mission to advance research, education, and policies around AI for the betterment of humanity.

A New Perspective on AI

Yoon: So in other words, HAI considers the social impact of AI and recognizes its responsibility toward human progress. I fully agree with this perspective and its direction. The development of AI-related technology is rapidly accelerating beyond expectations and is being applied across a wide range of fields, placing us in a continuous transitional period. Therefore, what you are suggesting through HAI is the significance of a human-centric approach during these times, in order to drive positive change.

Another significant aspect for me was that HAI was not solely developed by computer science engineers since its inception, and this remains true to this day. The entire group of scholars at Stanford has contributed to its establishment, and it now serves as a platform for academics, businesses, and governments to recognize the importance of their roles and foster communication among one another. This highlights the significance of participation and collaboration, which HAI places great emphasis on in addition to its mission of developing AI for the benefit of humanity. What do you believe are the essential components needed to establish a cooperative system that involves academics, businesses, and governments? For example, organizing events like the ACM FAccT[4] is crucial as it provides an opportunity for individuals to engage in discussions and exchange ideas.

Li: Indeed, that is a great and essential question. As previously mentioned, I would like to reiterate that HAI's AI framework was established with the belief that it requires an interdisciplinary approach. In other words, we cannot tackle it alone, whether it be solely computer scientists or any particular field. When engaging in any endeavor, be it research, education, policymaking, or social contribution, we always prioritize considering three fundamental perspectives.

PERSPECTIVE #1: BEYOND ENGINEERING AND INTO SOCIOLOGY AND PHILOSOPHY

Li: First, we must conduct research into the impact that AI has on humans and society by integrating a diverse range of fields of study. We must recognize that AI is no longer confined to being a small, lower-tier subfield of computer science.

To achieve this, we must collaborate with universities, social scientists, and humanities scholars from all over the world. It is crucial to gain a comprehensive understanding of how AI will affect us and our societies to predict the various ripples it may create. Ultimately, comprehending the impact of AI on humans and society, we must recognize at a fundamental level that AI has evolved beyond being a small, lower-tier subfield of computer science.

PERSPECTIVE #2: AI IS *AUGMENTING,* NOT *REPLACING,* HUMANITY

Li: Second, it concerns people's perception of AI. The word that most often comes up related to AI is "replace." But AI is not intended to replace us; rather, it has the potential to enhance humanity, which presents a significant opportunity for progress. This is an important matter that we should discuss with economists or policymakers.

As someone who has been working in the healthcare field for ten years, this is particularly meaningful to me. In healthcare, AI can be combined with other technologies to improve doctors' diagnostic efficiency, enhance patient safety, and reduce the stress and fatigue of doctors and nurses. It can also provide better protection for the elderly, allowing their family members who care for them to have more breathing space.

Furthermore, AI can aid in the development of new medicines and vaccines and find cures for rare diseases. It has enormous potential to benefit humanity. It is important to remember that AI and machine learning can improve lives and increase labor productivity. In summary, the second core value of HAI's AI framework is that we should strive to research and educate AI to enhance and augment human lives.

PERSPECTIVE #3: AI TECHNOLOGY ADVANCES WHEN BARRIERS BETWEEN FIELDS CRUMBLE

Li: The third perspective is grounded in the recognition that the current level of AI technology falls short of its potential to assist humans. AI technology is still prone to instability, limitations, and difficulty in explanation. It is essential to acknowledge that no matter how much the infrastructure of AI advances, it must still adhere to human nature. Therefore, we must integrate fields that explore human nature, such as neuroscience, cognitive science, and psychology, into AI tech development. In other words, the third perspective in HAI's framework is that future AI technology should draw inspiration from human neurology and cognitive science.

Yoon: When dealing with AI, it is crucial not to overlook the interdisciplinary perspectives that span multiple fields of study, the trust in human-centered AI, and the drive to enhance future AI technology with human nature. These core perspectives were established through interdisciplinary discussions that overcame differences in opinions from individuals who actively worked under HAI's values, making them feel more grounded.

Since there are no concrete answers to questions about AI, many opinions can cause discomfort and chaos. Polarized ideas can also lead to divisions, which may explain why some people avoid debates on

AI. However, if we establish a clear and common value of developing AI to advance humanity and share our talks in this direction, we can more freely exchange our ideas. By having a clear goal for our discussions, we can carefully listen to various perspectives instead of wasting energy refuting them, leading to more positive results. When barriers between fields crumble, we can achieve the goal of building human helping AI.

HAI: Stanford's Human-Centered AI Institute

AI technology has penetrated deeply into our daily lives, impacting our economy, healthcare, education, and politics and posing significant ethical questions. Stanford University's Human-Centered AI Institute (HAI) was established in March 2019 with a mission to develop AI technology that benefits humanity. As of 2022, the institute is led by ten professors from diverse fields, with Computer Science Professor Fei-Fei Li and Humanities Professor John Etchemendy serving as co-directors.

The advisory committee is composed of businesspeople, politicians, and other academic and social contributors, including former US secretary of state Condoleezza Rice, former Google CEO Eric Schmidt, and co-founder of Yahoo, Jerry Yang. HAI is committed to advancing humanity's well-being through AI technology and does not align with specific political parties or factions. The institute's team, which includes dozens of professors and employees from diverse backgrounds, actively invites renowned scientists, educators, social justice advocates, legal and policy experts, artists, and representatives from all academic fields, public and private sectors, and nonprofit organizations to collaborate and work together with HAI's activities to build a better future with AI.

Their main activities can be broadly divided into three parts. First, from a research and development perspective, HAI jointly supervises the following initiatives led by Stanford professors from various fields:

1. Develop AI that is inspired by human intelligence.

2. Research, predict, and map the effects of AI technology on humans and society.

3. Design and develop programs that utilize AI to enhance human abilities.

Second, from an educational standpoint, HAI helps students and leaders of various groups at Stanford learn the fundamentals of AI and diverse perspectives related to it. Lastly, the institute hosts regional and national events to facilitate the implementation of practical policies related to AI.[5]

Government's Role in the Age of AI: Between Regulation and Innovation

Yoon: We have discussed the fundamental AI framework and its three perspectives as a basis for creating human-centered AI. Moving on to our second topic, we will delve into the role of the government in the age of AI. As with our previous topic, it is crucial and indispensable to explore this area.

The rapid advancement of AI technologies is causing governments, which should be facilitating innovation and managing potential risks with their systems, to fall behind. However, to promote continuous growth focused on human-centered AI, guidelines related to innovation, human training, fair competition, technology development and usage, and more are necessary.

I believe that now is the time for governments to take an active leadership role in the discussion of building human-centered AI. In the US, the passing of the National Defense Authorization Act for Fiscal Year 2021[6] included some interesting additions, such as the creation of the National Artificial Intelligence Initiative Office (NAIIO) under the White House's Office of Science and Technology Policy. This emphasizes the intervention and action of the US Department of Defense to ensure AI development is ethical and socially responsible.

The act also calls for the National Institute of Standards and Technology to develop an "AI Risk Management Framework"[7] to manage the potential risks from the development and utilization of AI-related products, services, and systems. Compared to the past, it is encouraging and extraordinary that the government has taken the initiative to lay out systems and policies for continuous AI technology development based on social responsibility and ethics. Unfortunately, there are still some countries where AI-related policies and guidelines were rushed without considering multidisciplinary discussions and cooperations.

Last year, you collaborated extensively with the US government and made a significant contribution to the passing of AI-related legislation. Your leadership was crucial in establishing AI research and development strategies[8] within the US government. I remember you stressing the significance of the government's participation and support in the field of AI. I am curious about your thoughts on why the government's participation and support are essential for AI development, what role the government should play in promoting continuous AI advancement, and what should be at the core of

AI-related policies. Can you please share your perspective on these questions? Should governments aim for regulation or innovation?

The Government's Role in Encouraging Innovation

Li: Those are excellent questions, but before I can provide a thorough answer, I would like to address two questions that you raised. First, what did HAI do to encourage the participation of the government, and second, what should be the government's role in the AI industry— innovation or regulation? These two questions are closely intertwined, and understanding the relationship between them is crucial to understanding the government's role in AI development.

Allow me to give you a brief overview of the joint activities between HAI and the government in 2020. HAI led the effort in proposing the National AI Research Resource Task Force Act 8,[9] which became the legal basis for the establishment of the National Research Cloud 7 (NRC)[10] in collaboration with universities and businesses across the country. The NRC will make it possible for AI researchers in both academic and public domains to freely access relevant data and help increase investment and support toward AI-related research. This process demonstrated the government's recognition of the need for sufficient data and computing infrastructure for AI development.

Of course, breakthroughs in AI technology require not only brilliant algorithms but also sufficient data and computing infrastructure. The US has been at the forefront of the computer revolution, leading the development of bioengineering and pioneering other fields. This is the result of a healthy ecosystem where the government played a vital role in supporting basic science research in public institutions.

Entrepreneurship, especially in highly efficient industries like IT, was flourishing, and appropriate government subsidies and support contributed to its success. Now that we've entered the age of AI, trends in technology are changing rapidly, which allow private companies with exclusive data resources and computational infrastructure to quickly become leaders of certain technologies. Furthermore, government support is crucial for educating people, as the success of AI development depends on a well-trained workforce. Where do these people receive education? Universities such as Stanford or state universities play a significant role in educating future generations, and it is the government's responsibility to support and invigorate educational infrastructures. For steady growth, we must maintain a healthy ecosystem where everyone is mixed together.

Government Acted as a Catalyst When New Technologies Emerged

Li: The next part addresses the second question, namely, the role of the government in innovation versus regulation. Is innovation truly sustained when government intervention is not involved in the development process? I agree that the US has a healthy ecosystem for innovation, with a well-formed free market economy and an established entrepreneurial environment. I also believe that the government is playing a crucial role in that ecosystem, and the same holds true for AI development.

To illustrate, we can reflect on the early days of the internet. It is undeniable that federal government funding played a pivotal role in its creation. Various affirmative actions and subsidies acted as catalysts in leading digital innovation in the US for fifty years after World War II. I don't think even the entrepreneurs in Silicon Valley are well aware of this. In other words, our innovation would not have been possible without the active support from the federal government. Therefore, in this sense, the government's role has been exceedingly positive to this day.

Yoon: I see. The evolution of the internet provides a valuable point of reference for our current progress in AI. The innovation that occurred in the internet sector about thirty years ago, fueled by international investments and active government support, serves as a model for technological advancement. In recent years, the US government has significantly increased funding for AI research in the public sector [as reflected in Table 3], which is reminiscent of the government's efforts in advancing the internet. Therefore, we can anticipate that the government's support will have positive effects on the advancement of AI, similar to how it played a critical role in the evolution of the internet.

Up until now, we've focused on the government's role in providing funding and resources to build infrastructure and support innovation. However, there is another crucial function that the government can fulfill. Let's dive into that next.

A Guardrail to Prevent Technological Derailment

Li: Of course. The government has a crucial role to play as a regulator in the development of AI technology. As we have discussed earlier, the influence of AI is rapidly growing, and with this growth comes the potential for harm, privacy breaches, and biased algorithms. To mitigate these risks, the government must establish regulations that create a safety net for individuals and ensure fair competition in the commercialization of AI technology. In other words, the government must provide the necessary foundation for AI development.

Yoon: I agree. Cutting-edge technologies, such as AI, can be difficult to regulate once they gain momentum, making review and regulation all the more important to ensure that the technology is going in the intended direction and benefiting mankind. While some computer scientists and engineers may view government intervention and regulation as obstacles to innovation, it's important to reiterate that regulations are not meant to hinder development or innovation, but rather to ensure that AI's development stays on track toward its intended purpose. Unfortunately, there are cases where people outright oppose government regulations without acknowledging the legal limits behind them. While constructive debates from various perspectives are healthy, it's important to educate people, including developers and engineers, about the government's role in regulating AI and how it can help keep the innovation on the right path.

TOTAL AI-RELATED US FEDERAL CONTRACT SPENDING[11]

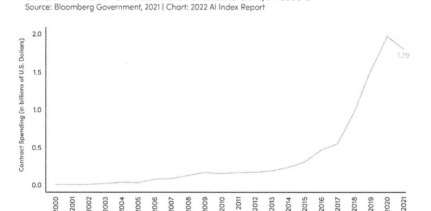

U.S. GOVERNMENT TOTAL CONTRACT SPENDING on AI, FY 2000-21
Source: Bloomberg Government, 2021 | Chart: 2022 AI Index Report

Daniel Zhang, Jack Clark, and Ray Perrault, "AI Index," Stanford University, May 16, 2022, https://aiindex.stanford.edu/.

Global Cooperation and Participation for All of Humanity

Li: Yes, and in that sense, I hope the discussions we're having here help to spread awareness of the government's role in AI's innovation. Lastly, another role of the government I'd like to emphasize is the collaboration with various nations, institutions, and groups with the same goals. By promoting global unity and collaboration, we can work together to ensure that AI is developed and used in a way that benefits humanity as a whole, without discrimination or violation of fundamental human rights. This is an important responsibility of the government and one that should not be taken lightly.

Yoon: Given the global impact of modern technologies such as AI, it is imperative for countries to collaborate with each other. According to the 2022 AI Index by HAI,[12] only one out of twenty-five countries had passed any AI-related bills in 2016. However, this number increased to eighteen in just five years, presumably due to the efforts and interactions among governments. As you highlighted, these international partnerships are crucial in developing human-centered AI technology. This requires frequent and ongoing discussions and exchanges among nations to share frameworks based on common values.

Human Evolution in the Age of AI

Yoon: As you mentioned in your reflection, we can observe the remarkable progress of AI since its early days. Facial recognition, for instance, swiftly became an integral part of our daily lives. What began as a simple function of identifying human faces has now evolved to become an intricate tool in apps that can verify a user's identity. It is also used to easily unlock phones through facial recognition and to automatically tag and categorize portrait photos on networks.

While such advanced technology is undoubtedly convenient, it is concerning that the use of these technologies does not always end there. Facial recognition can now be integrated into consumer drones to spy on neighbors or to track peaceful protestors by dictatorial governments using satellites. Advanced technologies with limitless potential for the benefit of humanity can also evoke fear and anxiety because we cannot be certain about their application.

Many have predicted that AI robots will replace most tasks currently performed by humans. Similarly, concerns about the replacement of humans by AI are also in the same context. AI ethics has recently become a major issue, such as the bias of facial recognition technology, invasion of privacy, fairness, and more. It is now essential to discuss and act upon the capabilities we must promote to live in the fast-paced, ever-changing era of AI and how we can prepare for future generations.

The third topic I would like to discuss with Dr. Li is the potential issues related to artificial intelligence. As an expert in AI, I would like to know your thoughts on the concerns raised by some regarding AI's capacity to replace human labor. We are already in the age of AI, and it is crucial for us to be aware of the developments in this field. What do you think are the key areas that humanity should focus on in this rapidly changing world?

NEW GENERATION FACING THE AGE OF AI

Li: I believe that all the points you made are important and should be taken into consideration. However, I think the most crucial aspect in resolving these issues is the role of people. If I had to pick just one thing from all the aspects you mentioned, I would choose the upcoming generations who will live in the AI era. It is important to recognize that the issues we face today may take more than one generation to address.

When I refer to the new generations, I am including individuals such as politicians, engineers, and other workers who have a deep understanding of both computer engineering and humanities. However, I do not believe that software developers should necessarily obtain doctorates in both fields.

An analogy can be drawn from the process of learning to drive in the US, where teenagers typically begin at around sixteen years old. They learn the basics of the car's engine, the mechanics of how it moves, the rules of the road, and their responsibilities as a driver. This analogy highlights the need for future generations, regardless of their profession, to receive a comprehensive education that strikes a balance between technology and humanity.

It is essential for politicians to understand how the internet industry operates, and technicians should be held accountable if their machines cause critically dangerous situations. I once had a conversation with a software developer in Silicon Valley who was in tears as they shared their experience. They expressed their desire to use technology for good but were distressed that their education, from kindergarten to university, had not included any classes on AI ethics or related topics. It was not until they became a software developer that they first encountered these issues, leaving them feeling ill-equipped to navigate the complex ethical considerations of their field. This is the reality of computer engineering. It is a daunting reality.

We must strive to ensure that future generations are not faced with similar situations. This is why I remind myself daily, upon waking, of HAI's mission and the importance of our work. I keep this in mind every day, knowing that our efforts today will have a significant impact on the world of tomorrow.

Yoon: I wholeheartedly agree with what you've said. It is our responsibility to nurture the resilience of the next generation, equipping them with the skills they need to navigate the challenges brought about by technological advancements. Achieving sustained success across generations cannot be achieved through short-term development alone. It requires an education system that can consistently train and prepare future talent.

To maintain a healthy and stable technological ecosystem, governments must prioritize investing in basic science research and supporting individuals who are looking to venture into businesses and society. This will require collaboration between education, private companies, nonprofits, and all those who are advancing the age of AI, to ensure that the next generation is nurtured in the right direction—one that balances the mastery of advanced technology with a focus on humane values. It's inspiring that HAI is located at Stanford, where their mission is to advance humanity. By gathering the core players of the current age to discuss and spread the perspective of a human-centered AI, HAI is contributing to a crucial educational process for our future generations.

Who Has Responsibility and Authority

Yoon: Discussions about the advancement of human society cannot neglect the issue of inequality. I'm interested to hear your thoughts on automation, which has been criticized for its potential to benefit those who possess the related technology or data, while taking away opportunities from those who do not.

Li: AI is not the first powerful tool humanity has ever developed. Throughout history, influential man-made tools have possessed both positive and negative aspects. Take electricity as an example: It brought light and heat to more people than any other invention in history. However, it also contributed to inequality in the market economy.

AI will likely follow a similar path. AI has the potential to deliver significant benefits to humanity, accessible by all. Yet, it can also enable individuals with the right business models and data to monopolize wealth and power. As such, AI technology alone does not determine right or wrong; it is merely a part of the equation. Policies, established regulations, and societal participation related to AI complete the equation.

My primary concern is the potential neglect of AI. We have learned from the internet that technology itself is not inherently evil. Evil arises when people act irresponsibly, when sound policies are absent, and when the balance between proper regulation falters. The value of machines is determined by humans, meaning our values are reflected upon them. To illustrate this, let's consider COVID-19, as I am involved in healthcare projects.

Remote medical treatment technology advanced rapidly during the pandemic. Thanks to these services, elderly or chronically ill individuals who could not leave their homes were able to receive medication and care. I, too, had video calls with my doctor when feeling unwell—a conversation we never previously had remotely. Remote medical services demonstrate the potential for progress in healthcare. However, we must consistently ensure that these systems are accessible to all and do not operate unfairly or unjustly. Ultimately, we must remember that human decision-making drives all system operations and processes. If we lose sight of this, inequality may arise in various forms. To prevent this and ensure fairness, we must maintain an interdisciplinary approach—from policymaking to education—related to AI, thus preserving checks and balances.

Yoon: Yes, you made a very good point. It is crucial for future leaders to responsibly utilize AI as a powerful tool. Educators play a vital role in training the next generation to recognize not only the benefits of technological advancements but also the potential dangers, ensuring they prioritize human values throughout development processes.

Li: I agree that business leaders like yourself have significant roles in guiding future generations. We are all learners and educators in our own domains, continuously acquiring knowledge while leading and sharing our convictions with others. Industry leaders are as essential as those in academia, and all societal stakeholders must work together to address the challenges and opportunities presented by AI. In that sense, I would like to hear your perspective as an entrepreneur. You've been leading debates and suggesting directions on AI ethics. And you've also been sending over advice and support as an advisor to HAI. What do you think is the role of businesspeople in dampening the potential threats of AI technology?

Yoon: I don't think I can give an answer right now. This is why I asked about your thoughts and perspectives. Businesses like ours are often expected to create something fair without a clear definition of what fairness means in our society. However, I believe that acknowledging the need for responsibility is a good starting point. Fairness may not be universally perceived in the same way by different groups of people. Recognizing the limitations of certain perspectives and examining related issues from multiple angles can be beneficial.

Businesspeople or engineers should not solely determine what is fair just based on their own viewpoints. Instead, we need to understand AI technology from various perspectives and engage in open dialogue with trailblazing researchers like you. I am grateful for the opportunity to hear your sincere and articulate thoughts, and I look forward to continuing this important conversation on the ethical development and application of AI. The age of AI is already upon us. What should humanity pay attention to in this ever-changing world?

Cooperation beyond Nations and Cultures

Yoon: Our discussion began by addressing diverse perspectives on AI, transitioning to the balance between innovation and regulation in government roles and then focusing on new generations living in the age of AI. We now arrive at our final question. We live in a world encompassing a myriad of viewpoints, experiencing significant cultural clashes. The intensity of cultural collisions and differences in perspectives among interest groups, generations, and political parties has never been greater.

Even if the government provides guidelines and policies related to AI technologies, they cannot control every line of code entered by developers. In this case, biases in AI algorithms arise not from the technology itself but from the prejudices ingrained in the society the engineer belongs to. However, recognizing these prejudices, agreeing upon common values, finding shared values to protect, seeking harmony between highly polarized perspectives, and preserving our dignity as human beings through various challenges are tasks that even the most skilled teachers or politicians struggle to accomplish.

Do you believe that leaders in the field of AI can work together to succeed in such a difficult task? How can leaders of AI technologies appropriately address these issues? While I still don't have answers to these questions, I am eager to explore possible paths in conversation.

Seeking Breakthroughs from Various Perspectives

Li: These are indeed fundamental and profound questions for our current age. I cannot claim to have definitive answers either. I believe that you, me, and everyone at HAI are collectively engaged in the process of seeking answers. HAI is making considerable efforts to create a platform.

As a nonprofit educational institute, we occupy a relatively neutral position, which allows us to freely ask open questions like "What does it mean to design AI?" or "What would be the endpoint if we want AI to be fair and humane?" on this platform. To advance debates, we must consider laws, regulations, social media, social movements, and other tools of our era. Ultimately, however, we need a forum where interdisciplinary groups can talk and experiment together.

For example, facial recognition AI is an influential technology that has raised concerns about biased data being used to disadvantage certain individuals in legal proceedings, financial systems, and medical systems, among other areas. This has attracted significant attention and concern, with different generations and diverse perspectives weighing in on the technology. In response, HAI organized a forum in 2019[13] that brought together representatives from multiple groups, including American Civil Liberties Union (ACLU)[14] lawyers, the Algorithmic Justice League (AJL),[15] federal groups like the National Institute of Standard and Technology (NIST),[16] regional and state policymakers, legal experts, essentialism and realism scholars, and computer scientists.

Of course, one or two such discussions are not enough. Bringing together a diverse range of people to share their perspectives and listen to each other is an important first step. It's not easy for interdisciplinary individuals to convene and engage in conversation, but without such gatherings, we cannot identify potential resolutions or conduct necessary experiments. By assembling, participants can understand others' positions and reach a consensus, which will aid in the discovery of more universally beneficial solutions. HAI aims to establish a platform that facilitates these conversations.

We also support related projects or research beyond merely creating the platform. At HAI, legal experts and computer engineers carry out these tasks. We also participate in activities to directly promote policies or assist policymakers. Furthermore, we host events that bring together individuals from various social groups, including council members, media journalists, and corporate board members. These interdisciplinary participants engage in numerous discussions, creating diverse content that enriches their conversations. The content generated during this process is then used as educational material to foster cooperation between different groups. While I may not have provided definitive answers to your questions, this is the methodology I believe in: inviting and finding answers together with interdisciplinary people, leaders from various generations, thinkers, and diverse members of our society.

Yoon: Yes, I agree that holding forums where interdisciplinary people can understand one another is an important first step in finding answers. However, it will be difficult for them to start talking if the participants use their own terms. It is a great task in itself to gather people of various groups to a single location, but it will be equally difficult for them to communicate fluidly using the same terms.

Li: You're right. For that to work, people must gather and engage in conversations repeatedly. Your experience with the healthcare project illustrates this point well. It took years for you to learn the terminology used by doctors and understand their discussions, and this effort continues to this day.

Becoming One beyond Borders and Boundaries

Li: I also have one last question for you. You are from Asia and have international experiences. Sometimes the focus can be overly centered on the US, so I'm curious about international perspectives. How do you think we should establish horizontal international relationships regarding technologies? What processes do you think we should have to spread not just trust in certain technologies but also the idea of using them ethically, fairly, and benevolently?

Yoon: I believe it ultimately comes down to establishing solid foundations. To horizontally disseminate trust in technologies among global leaders, we should begin with the most fundamental principles. For example, personal data ownership and usage are viewed differently across various countries, with some systems not even recognizing individual data ownership. Without a common perspective on such basic matters, finding common ground on the principles and necessary foundations for AI technologies becomes quite challenging. Therefore, we need a holistic approach that takes into account legalism, democracy, trust in humanity, freedom, and technological advancements.

If we are to seek for a better future encompassing important perspectives from various angles, we must come together. There is no alternative but to consistently collaborate for the advancement of human civilization. Time is limited, and as it passes, we will inevitably make mistakes. However, these experiences will offer us the opportunity to recognize not only the benefits of AI but also its drawbacks and potential harms.

Li: I agree. As someone working in a university research lab, I am greatly inspired by business leaders like you who possess a clear sense of responsibility for both the present and future. We have a long journey ahead of us, and I am grateful to have a friend, advisor, and supporter like you. Our personal conversations and collaborative projects have been enlightening and heartwarming experiences. I look forward to the continuation of our partnership.

Yoon: Thank you so much. During our conversation, I could feel my thoughts becoming clearer and more distinct. The future world, reshaped by AI technology with limitless potential, is filled with uncertainty. However, we must move forward with trust instead of fear in order to create a better future. What's crucial is that we work together, taking cautious and deliberate steps to avoid missteps along the way.

The blueprint proposed by pioneers like you, backed by years of research and experience, will serve as the foundation for a shared value. Governments, businesspeople, scholars, and individuals from all fields and generations should engage in ongoing discussions to find their roles in creating AI that benefits humanity. I believe this path will lead us toward technological innovation and human progress.

Supporting you and HAI, which are leading the way toward responsible AI development and positively influencing the world, is both an honor and a pleasure for me. Thank you for such a deep and meaningful discussion.

CHAPTER 2
Unavoidable Dilemma

*Answers to the Future Lie in Education
Combining Humanities and Engineering*

AI (Education) Framework × Rob Reich

Dr. Rob Reich is a political science professor at Stanford and associate director of HAI (Stanford's Human-Centered AI Institute). He teaches political science and also has classes in graduate school of education at Stanford. He is also the director of the Center for Ethics in Society and co-director of the Center on Philanthropy and Civil Society at Stanford. His work focuses on ethics, public policy, and technology.

He authored *Bridging Liberalism and Multiculturalism in American Education* and *Just Giving: Why Philanthropy Is Failing Democracy and How It Can Do Better*, which go over political science and education, and recently published *Digital Technology and Democratic Theory* and *System Error: Where Big Tech Went Wrong and How We Can Reboot*, which highlights ethics in democratic societies centered around technology.

Problems We've Never Faced, in Need of Collaborative Solutions

Incredible advancements brought on by AI technology have now become part of our daily lives. When we look at our smartphones, they instantly recognize our faces and unlock our devices. Meanwhile, as technologies like facial recognition and end-to-end encryption advance, we're faced with issues we've never even considered before, such as individual privacy conflicting with national security. We're at a point where the imagination of a technological utopia also gives rise to concerns about a dystopian future.

Debates on how AI technology will change our society, what problems it will bring, and how we should handle them are now inevitable. We must maintain focus and discuss together the fact that technology is there to advance humanity and, in a world of clashing values, create a balanced perspective framework to establish ethics in this new age of coexistence with AI.

This process cannot be achieved with the perspectives of just AI experts or philosophers and ethicists. Nor can it be done independently with government regulations or policies. Scholars worldwide are seeking answers through participation, discussion, and exchange between many fields covering humanities, social sciences, engineering, and more to build an ethical framework. To properly handle rapidly advancing AI technology, related information is being shared publicly, while universities are using AI ethics curriculums to try to change the world from a more long-term perspective, covering both current and future generations.

In this chapter, I interact with Dr. Rob Reich of Stanford University, who quickly acknowledged the influence of AI and began teaching classes combining engineering and ethics. As an active political scientist in the age of AI, Dr. Reich has been emphasizing the importance of an ethical framework and education to counter potential threats from AI. I will discuss with him the importance of ethical perspectives we must pay attention to as individuals living in a society coexisting with AI. This will also be a session where we gain insight into educational solutions to preserve and pass on these perspectives.

Humanities Scholars' Role in the Age of AI

Yoon: Greetings, Dr. Reich. Thank you for joining us today. The topic we're going to cover is something I am quite fond of. It involves closely observing the changes in societal order brought on by the rapid advancement of cutting-edge technologies and examining the choices we've made and will make as members of society living within the chaos of transition. I would like to start by discussing your recent book, *Digital Technology and Democratic Theory*. Based on theoretical perspectives of a democratic society, what do you think are the technological changes and issues we should pay attention to?

Reich: Nice to meet you. I am also delighted to have the opportunity to talk to you. One of the great things about working at Stanford's HAI is having the chance to engage with various people like you outside of the university. As a scholar, I learn a lot from these interactions.

31

Digital Technology and Democratic Theory is, in many ways, connected to the missions of HAI. Numerous digital tools, IT services, and platforms were born in labs at Stanford University. These technologies were then distributed to the world through Silicon Valley companies, leading to the digital revolution. When these digital technologies first began to spread, many believed that digital tools would advance democracy and contribute to the freedom of the entire human race. However, over the past four to five years, digital tools, platforms, and IT services have instead become means of surveillance and oppression, distorting or hiding the truth and spreading misinformation by undermining traditional media and journalism. We witnessed something opposite to what we had hoped for. As a result, forecasts on technology shifted from optimism to solid pessimism.

In reality, digital tools can be both good and bad for our world. Now, we must work diligently to consider the interaction between powerful tools, platforms, IT services and democracy, democratic ideals, and the digital economy. In this perspective, *Digital Technology and Democratic Theory* aims to provide a more realistic and mature viewpoint for setting future research agendas. That is why this book was written in collaboration with scholars from various fields, including philosophers, economists, computer scientists, political scientists, and psychologists. We shared and discussed a wide range of subjects without leaning toward the extreme. I hope people from different fields have the chance to read this book so that we can progress constructively and contemplate within their own fields how to suggest methods to mitigate potential damages from the digital revolution.

Yoon: I can certainly see from what you've said how both the theme of the book and its writing process are closely related to the goals of HAI. After all, HAI is an institute that combines knowledge from many fields and seeks common wisdom on how humanity can coexist with AI technology. In my discussion with Dr. Fei-Fei Li, she emphasized the establishment of a platform that allows these conversations to thrive through HAI, and it appears that *Digital Technology and Democratic Theory* is an example of such a platform. It includes the opinions of computer scientists who are deeply involved with digital technologies, political scientists like yourself, and those of economists, psychologists, and other scholars from different fields. As a result, readers will learn about various perspectives from this book, which, in turn, will provide them with an opportunity to voice their own concerns and ideas.

Reich: That is true.

Ethicist Participation from the Beginning

Yoon: It hasn't been long since people with various perspectives began to comfortably exchange and consolidate their ideas. Computer engineering and AI, in particular, seemed quite distant from philosophy and social sciences. However, the influence of AI has expanded each year, now extending beyond the industry into politics, economy, and culture.

Sharing your experiences as a political scientist participating in HAI will be very insightful. How did you come to join HAI, which is primarily focused on computer science and AI? Additionally, I'm curious about your perspective as a political scientist on participating in HAI's process of building a better future and what you consider to be the most important aspect of your involvement.

Reich: I'm going to be frank about this. I've been a political science professor at Stanford for about twenty years, and in the last decade, the number of students majoring in computer sciences, particularly AI, has increased dramatically. I became curious about what made these fields so appealing to so many people. Although majoring in computer sciences can lead to higher salaries, the sheer increase in numbers indicated a significant change driven by the students themselves.

I wanted to understand what was happening within the computer science and AI departments on campus, so I spent time there. That's when I had the opportunity to meet Dr. Fei-Fei Li and other HAI leaders. After numerous conversations, I found myself in agreement with HAI's core principles. When it comes to considering the future of AI, the roles of philosophers, ethicists, and sociologists should not begin only after AI researchers invent something in the lab or when a company commercializes related technologies.

In other words, contemplating the impact of inventions on the world after they have already been introduced by developers is too late. I wanted to be in the lab with AI researchers, discussing ethical issues, democratic thinking, and interactions with democratic systems from the design or early stages of ideas. The same goes for my social science colleagues. Unlike other institutions, HAI has gathered renowned social scientists and philosophers. HAI leverages this capability to assemble skilled scholars from various fields, ultimately helping to build a better future.

Yoon: Thank you for the honest reply. I am also well aware of the immense influence and significance of philosophy, ethics, and social sciences. However, it's true that unlike the field of AI today, social sciences such as ethics and philosophy haven't been actively involved in the early stages of mechanical or electrical engineering. Why do you think there are different approaches to other science fields in comparison to AI? Why does AI specifically need the participation of social sciences?

Leich: Well, I partially agree with your perspective and concede that the emphasis on ethical participation is not exclusive to the field of AI; it is true that areas such as bioethics, medical research, and drug development have a long history of engaging with ethics and philosophy. Institutions like the Institutional Review Board (IRB) and the US Food and Drug Administration exemplify the connections between these fields and ethics, covering a wide range of societal aspects. Medical schools require ethics classes, and hospitals generally have their own ethics commissions. I think there are some similarities in the history of engineering as well.

That being said, as you pointed out, there are unique aspects to the field of AI. The most significant distinction between AI and other engineering technologies is that computer engineering as an academic field did not exist before the 1950s. It is a relatively new discipline that has rapidly gained prominence across the globe. The dual revolutions[17] of AI and bioengineering, along with targeted gene replacement, are arguably the most groundbreaking discoveries or revolutions of the twenty-first century.

For instance, the combination of gene replacement technology and AI could have devastating consequences if not properly managed. This is why universities and the global community need to closely monitor these new discoveries and revolutions to ensure they solely benefit humanity. Furthermore, universities must strive to prevent various research endeavors, discoveries, and inventions from being unduly influenced by specific companies or political ideologies. Instead, they should pursue the truth and contribute to building a better world. In this sense, I believe universities are facing a critical juncture in the twenty-first century, where these powerful revolutions could potentially determine the fate of humanity.

Yoon: I don't think it's easy to recognize that computer engineering and AI are closer to bioengineering or medical science than other fields of engineering. Medical practices usually treat people through direct contact, so AI feels like it's at the opposite spectrum. That is why I am greatly intrigued by what you've said. Could you give more detail on this?

Reich: Yes, you're absolutely right. Although AI doesn't involve direct physical interaction with the human body, it can still have significant effects on our emotional well-being and overall quality of life. However, think about the following example. Six to seven years ago, AI developers at Facebook experimented on the influence of changes in newsfeed algorithm on the emotions of their users. They tried giving positive and familiar contents more priority and lowering them for depressing ones or vice versa. Newsfeeds do not directly make contact with our bodies, but this was an experiment to specifically investigate their effects on human well-being.[18]

I believe this also applies to various AIs such as Siri or Alexa. AI can have influence over human emotions and can affect our information environments, which makes them closely related to the quality of life. Of course, we can't say that they are identical to drug experiments injecting substances directly into veins, but AI also has a chance to cause the same issues or suspicions that drug experiments can result in.

Technology in Ethical Dilemma Needs a Frame for Balance

Yoon: The example you gave makes it very clear. Indeed, AI has the potential to influence us psychologically and emotionally, which can subsequently impact our physical well-being. Given its widespread effects, it is crucial to incorporate philosophical and ethical considerations into AI research and development to ensure that these technologies are developed responsibly and with the greater good in mind. However, there are those who argue against the involvement of philosophical, ethical, and political perspectives in AI research, claiming that regulations and ethical guidelines may stifle innovation. What are your thoughts on this?

Reich: I hope these people can think beyond such stereotypes soon. Many people think that when ethicists like myself work with engineers or scientists, I tell them to slow down or suggest stopping something or ask questions like "Is that really a good idea?" bothering their work. They think this is the role of ethicists and also believe philosophers hinder the progress of companies. Of course, this is sometimes the case, but I don't think that's generally the role of philosophers.

It's like saying something like "I am the expert on ethics, so you do what you do, and I will decide if your work is right or wrong." Ethics considerations are unavoidable for everyone. We all have ethical standards, and we face ethical issues in our daily lives and at work. I don't want to be an expert who decides on what is right or wrong, but rather someone who provides frameworks for ethical perspectives when values collide in our daily lives, society, and especially at the forefront of technological advancements.

Let me give you a specific example. Let's take a look at surveillance technologies on our smartphones and cookies, which collect various data from your web browsers. Some people think these are privacy violations and that the internet itself intrudes upon our privacy.

Also, there are people who argue that they should use apps like Signal or WhatsApp with end-to-end encryption[19] to prevent the government or companies from censoring their messages. This could be more effective in protecting privacy. But the government could counterargue with something like this, "What if terrorists use these platforms to plan out acts of terrorism?" or "If sexual offenders use them to commit crimes, how do we deal with them?" As such, privacy and safety, or privacy and security, contradict each other.

This philosophical challenge is interesting not just for philosophers but for everybody. Whether you're a developer creating surveillance tools within a company or a policymaker on public regulations, you will face contradictions between protecting the privacy of individuals or their safety under national security. How do you balance this? "Trusting your gut" is not the method to go forward. It is at the moment of discussing such balance where philosophers can jump in.

Yoon: Yes, I have more clarity now. As you've mentioned, we're faced with daily instances where we can experience harm or losses from commercialized AI technologies. We've discovered that tools utilizing AI can exhibit prejudice or discriminatory tendencies, and there's growing distrust in social network services using algorithms. Can you share your thoughts as an ethicist?

Reich: There are indeed many concerns and questions on this subject. First, in the absence of a common ethical standard, tools using powerful facial recognition could trigger a race to the bottom. Regardless of what companies do or what their customers want, we don't want the facial recognition tech in our smartphones to advance to a point where it becomes something that recognizes every single person walking down the street.

Questioning the use of facial recognition on drones or in public safety or military situations follows the same logic. Today, AI is spreading through automated systems and robots and technologies that replace labor. So, an important question we must ask right now is, "In the eve of the AI revolution, what is the future of work?"

This leads to further questions such as "If companies earn profit by using automation tools and systems, what responsibilities should they have for the loss of workplaces caused by them?" "Should the government take responsibility, or should it be individual business people or companies?" "Can governments calculate the proper amount of robot tax?" and "Were there any discussions on using AI tools to augment human abilities instead of replacing them?"

We should also consider these points. When people talk about tools for diagnosing cancer, they often think about various types of computer-based visual devices that can instantly diagnose early-stage skin cancer or help enhance human vision to find them. But do we really want a world where machines diagnose cancer for us? Or do we want doctors to participate in the process? Whichever way, I think that people generally want a harmony of mechanical and human intelligence in certain things. These are questions we should all think about. They're not something we should rely on just computer scientists or philosophers for. We should have a comprehensive approach to all parts of the process and think about them together.

Yoon: I agree. AI technology is bringing revolutionary changes at every moment, which, in turn, presents us with new questions in the face of all these changes. No single perspective or approach can be the definitive answer. This is a period where we must contribute our own perspectives to establish a common ethical basis for the new age. Whether you're an expert in AI or not, a great scholar or not, the perspectives of everyone living in this era of change are all meaningful. As such, we must responsibly express our opinions and recognize the importance of listening to one another.

Bioethics and Biomedical Ethics

The term "bioethics" was first coined by Van Rensselaer Potter in his 1970 published book *Bioethics: The Science of Survival.* According to Potter, bioethics was a new field that combines biological knowledge and knowledge of human value systems, with the mission of enhancing human dignity and value. In the late twentieth century, when this concept emerged and started to be discussed, there was an increase in the development of medical technology and bioengineering, such as artificial abortion, organ transplants, and artificial insemination. As a result, the discussion and academic research of bioethics mainly revolved around medical procedures.

The successful cloning of a mammal, Dolly the sheep, through somatic nucleus transfer at the Roslin Institute in the UK in 1996 and the stem cell manipulation incident by Professor Hwang Woosuk's team in Korea in 2005 sparked global discussion on bioethics. Cloning, especially concerns over the possibility of human cloning, made people realize that modern medical and bioengineering advancements require thorough ethical, legal, and social considerations.

This also ignited policy and system research on ethical foundations for bioengineering and healthcare. The incident concerning Professor Hwang Woosuk's team revealed ethical issues during the research process, such as the illegal sale of eggs and forced egg collection from female researchers. In addition, evidence to cover up the truth emerged from both the researchers and the related government agency, highlighting the lack of ethics in domestic academic research and the management or regulation system regarding research on humans. The Korean government revised the entirety of its Bioethics and Safety Act, which previously revolved only around medical procedures, in 2012.

This came into effect in 2013 to provide an institutional foundation. According to the revised act, ethics is defined as follows: First, bioethics encompasses the protection of the participating human subject and related ethics in research processes. This includes securing human dignity, gathering research subjects under proper procedures, protecting research subjects, securing private information of research subjects, and more during research processes. Second, the other aspect of bioethics is for research results to protect human dignity while commercializing and industrializing the results safely and ethically. This includes the prohibition of human cloning and gene testing related to intelligence, obesity, appearance, and more; sales of human eggs, sperms, and other reproductive cells; and limitations on surrogacy.[20]

Ethics Education for a Better Future

Yoon: We've discussed the potential threat of AI technologies and the importance of ethical consideration and discussion at early stages of ideation to counter these threats. As your participation in HAI as a political scientist shows, AI is no longer limited to computer science or engineering but has expanded beyond academia, involving a diverse range of people working together.

Considering the social changes up to this day, interdisciplinary efforts are becoming increasingly important. Providing an ethical framework for the stability of human society in these complex times of clashing values and persistently and continuously addressing new questions brought on by the age of AI are all in their early stages. Education, which is our second topic, is a crucial aspect in navigating these challenges. As a political scientist, you have dealt with political and ethical perspectives on modern technological civilization and emphasized educational approaches throughout your twenty-year career as a lecturer.

Determining the priority for education is essential when expanding efforts to equip individuals with the knowledge and standards needed to live fulfilling lives in the ever-changing age of AI. It raises questions such as whether consumers who experience the potential threats of AI firsthand should receive necessary information first or if the government and AI developers should be prioritized. AI has transcended the confines of computer science and engineering, spreading beyond academia and integrating perspectives from a variety of people working together.

The Sum of Everything We Can Do

Reich: I believe that all the parties you've mentioned are equally important. However, relying solely on consumers to address all the issues won't resolve anything. A one-on-one approach makes it difficult for us to employ broader collaborative methods. For instance, some people delete Facebook from their phones in frustration, thinking it's the only solution. Others argue that smartphone screens should be black and white to reduce distractions from notifications and claim this is an exercise of consumer power. If these are the only actions we take, they may ultimately be detrimental.

Companies, along with consumers, need to participate responsibly in the innovation process, while various nonprofit, charity, and civil groups must respond swiftly to different situations. Government intervention is also a key element.

Consider content censorship or regulation guidelines as an example. This is when we could go to the cinema without worrying too much. In the US, movies have their own ratings, such as General, PG, R, No Audiences, Parental Guidance Suggested Restricted Content, and so on. However, these ratings are not set by the US government; they are independent guidelines established by the film industry to provide necessary information to consumers. By consulting this information, consumers can decide whether a movie is suitable for their children or determine the required age for R-rated films.

The social media industry could adopt a similar cooperative approach. Facebook's Oversight Board[21] made an attempt, but it was limited to Facebook only. The importance of the Motion Picture Association example is that the majority of filmmakers follow this method, voluntarily submitting their content for ratings. This approach has a broader impact compared to individual consumer reactions and offers more flexibility than government-imposed legal regulations. My point is that we shouldn't limit ourselves to choosing between user decisions or government regulation; there are many options in between.

Yoon: I see. The case with the Motion Picture Association is inspiring. Individual consumer actions can be weak or fragmented, and expecting the government to create policies for every aspect of AI would result in extreme complexity and decision-making difficulties. Exploring various possibilities in between seems crucial. Facilitating the exchange of information through a process that connects individual perspectives would be advantageous. Moreover, engaging in open debates on diverse ethical viewpoints and having participants reach agreements and act on those agreed-upon matters can serve as an educational tool in itself. This process encourages understanding and collaboration, fostering a collective effort in navigating the ethical challenges brought about by AI technologies.

Reich: That's right. I'd like to add another point. One of the best aspects of teaching at my current university is having the opportunity to instruct nineteen-, twenty-, and twenty-one-year-olds. These students will likely hold positions of leadership and responsibility within the next ten years. Training the next generation of leaders with a foundation in human-centered AI and an ethical framework for the new era is a crucial transformation that universities are uniquely positioned to achieve, making it a long-term goal.

In reality, it is challenging for private companies to train individuals with a twenty-year horizon in mind. They often focus on achieving short-term goals, such as meeting their next quarterly sales targets. Fortunately, institutions like Stanford do not have quarterly sales reports or publicly traded stock. This freedom allows us to develop plans like "How can we educate students over the next ten years to make a positive impact on the world thirty years from now?"

Computer Science and Ethics Combined Classes

Yoon: I agree. When discussing education related to ethics in the age of AI, it is essential for universities to establish long-term goals in order to drive the changes of the future. Now, let's transition to our next topic regarding the ethics curriculum at Stanford. Could you provide an overview of this new curriculum? I'm eager to learn how ethics education is practically implemented in university courses and what early successes you have witnessed.

Reich: Thanks to support from individuals like yourself, HAI has been tackling various aspects of human-centered AI. I am particularly involved in two complementary activities. The first one is called Embedded EthiCS, in which we pair individuals trained in ethics, philosophy, political science, or other relevant fields with computer engineering professors. This approach enables collaborative curriculum development, rather than expecting professors without a philosophy background to integrate complex ethics content into their lectures.

Through this process, we are developing modules for core courses, allowing Stanford computer engineering students to gain broad exposure to ethical frameworks and relevant questions in AI and computer engineering. In other words, when students learn basic algorithm models, they also explore questions related to algorithm biases and fairness. Instead of directing them to separate philosophy classes, we incorporate ethics-related content into core computer engineering courses.

The second activity is a large introductory class that I personally participate in. This course features political scientists like myself, philosophers, social scientists, and computer engineering professors with experience in joint policy development. In fact, the most popular professor on our campus, Professor Mehran Sahami, has taught the computer science introductory course for the last ten years. He is a brilliant educator. Furthermore, Professor Sahami's class is the only one at Stanford where students can learn about engineering-related policies and philosophy in a single course. We designed this course to accommodate as many students as possible, with around three hundred students currently able to take it.

Naturally, computer science graduate students may not know how to assess philosophy papers, and philosophy graduate students might struggle with engineering papers, so our assistants have their work cut out for them. This is how we've been operating for the past three years. We also offer custom case study classes and evening classes for professionals in engineering fields. Based on our experiences, if more interdisciplinary classes are created at Stanford and other universities follow suit, both professors and students will be better equipped to address ethical and social issues arising from modern advanced technologies. I believe this will contribute to guiding the world in a better direction.

Yoon: That is such meaningful work. Providing young students at Stanford with these educational experiences is likely to result in significant social contributions. Also, I hear you have your curriculum on the web to allow more people to make use of it.

Reich: Yes, with certain conditions, anyone can access these materials. You can find them on Stanford's class web page.[22] Although some documents are excluded due to copyright, you can download all core course modules developed by Embedded EthiCS and files related to case study research. It is our hope that other universities and companies can freely use these resources, as they have been developed with the intention of benefiting everyone.

How to Apply Ethics Education to the World

Yoon: To progress research required by the current generation and to share the results with the world must be thrilling as both a scholar and an educator. I appreciate your interest in sharing the materials with companies as well.

As for our last topic, it's true that we don't always apply what we've learned in school to our everyday lives. While students who take the courses I've introduced will gain valuable academic experiences in shaping their ethical perspectives, there's no guarantee that they will directly apply these insights in their professional lives. I wonder, what are your expectations with how the students in your classes will implement their learnings in society?

Reich: A very good question. What I tell my students at the beginning of the course is that when a philosopher appears, people might shrug and think, "Here comes the ethics expert to tell us how to live properly." They're not exactly welcomed. I want to emphasize that there are at least three stages of mentality in ethics.

The first stage is ethics of the individual, but it's not a stage I find particularly interesting. Of course, if every person acted ethically, the world would be a better place. However, expecting everyone to be perfectly ethical is unrealistic, as humans are imperfect by nature. Thus, ethics alone cannot prevent all future wrongdoings. No one can be a perfect saint, and we all experience pain because of our individual flaws. So, if we are to prevent evil deeds of the future, ethics is not an ideal vaccine. There is no ethics shot to prevent the bad things we will face in the future.

I find stages two and three more interesting. Stage two is work ethics. In working environments, what are the rules and structures that bind everyone as one? A classic example is the Hippocratic Oath taken by doctors and medical workers, which includes "do no harm" and other procedural rules. In this sense, I believe we can develop broad and professional rules for AI scientists and experts, which would greatly contribute to the field.

The third and last stage is political and social ethics. How should we think about systems that form our behaviors? How can we design better systems? We need something to direct us to better paths. If an ethics expert tells people, "I've come to tell you how to become a better person," they may be met with skepticism. Most of us know we shouldn't lie, deceive, or steal. We all know that we should not use illegal substances like Olympic medalist Lance Armstrong to win the Tour de France or to deceive consumers by using the power of technology like Stanford dropout Elizabeth Holmes, founder of Theranos.

We're not opening ethics courses to tell you things like "Thou shalt not lie, deceive, or steal." If you haven't learned the basic principles of ethics before coming to Stanford, it's already too late. Instead, we need classes that discuss ethical measures from a systematic point of view, helping students navigate clashes of values and preparing them to be responsible democratic citizens.

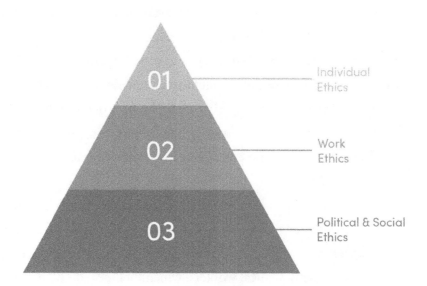

Yoon: It's a very important part. As you've said, it is a widespread misunderstanding that ethics education is outdated or merely about fundamental teachings of not doing bad things. The rapid development of AI continuously and broadly causes issues we've never faced before, making it inevitable for various values to clash with one another. What we believed to be right yesterday may change today, and we must question and filter out what is truly right within the turmoil of differences in environments and cultures. We have to remind ourselves that ethics education is necessary, not just in universities but for all to think about things with ethical perspectives and to learn from the choices we make in the age of AI.

My question in relation to this is regarding the third stage, political ethics. From the society's point of view, government policy is key. If the government is to spark innovation in AI but also guarantee an environment of ethical responsibility, which policy route should they take?

Reich: As this is not my field of expertise, and we don't have complete research conclusions on this topic, my answer is subject to change. There are general and special cases of government regulations concerning technology. From a general standpoint, scientific discoveries and technological innovations often outpace the government's regulations, as seen in the early days of nuclear energy, biomedical research, and the industrial revolution. Therefore, we shouldn't worry about policymakers not having the same level of knowledge as universities or industry pioneers in science and technology.

We're on a traditional route we've always taken, and with the current era and space advancing simultaneously, I'm not overly concerned about this aspect. That said, there are special cases where governments implement innovative regulations. For instance, Taiwan and the UK have experimented with a concept called the "Regulatory Sandbox." In this approach, companies develop new technologies or applications and then approach government regulatory agencies, stating they've developed a new product or service but are unsure of its potential impact. They propose basic regulations for their innovation, and the government agrees to release the product with the proposed regulations in place for a trial period of two to three years. Afterward, public policy professionals evaluate the effects of the initial regulations and make necessary adjustments. This process establishes a temporary regulatory environment and helps disseminate more information to society. It differs from the approach where companies operate freely until the government intervenes with regulations. Consequently, I'm interested in researching regulatory sandboxes further to understand their implementation in various countries and fields.

Yoon: I find this interesting because it is similar to what we have in the gaming industry. The gaming industry is known to independently provide regulations because, due to the content's characteristics, policymakers cannot go through them thoroughly. The US has the ESRB (Entertainment Software Ratings Board), which gives ratings to games and provides related information. Europe also has its counterpart, PEGI (Pan-European Game Information), which provides game rating standards, and Japan has a system for developers to voluntarily regulate themselves. Many gaming companies are proud to be part of a voluntarily regulated industry, which seems to share the characteristics of what you've just explained.

And now it is time for the last question. I see you have an optimistic attitude toward AI. And it coincides with HAI's stance, which considers AI as a necessary technology and tool to solve many social issues revolving around inclusion, diversity, equity, and many more aspects. Is there a specific reason why you think AI is a tool that can help humanity move in a better direction?

Reich: To answer this, I think we can compare AI to the growth cycle of humans. We could say that AI is in its teens. AI has a lot of strength and, like teenagers who are uncontrollable, seems to want to independently make use of its new strength, not listening to its parents. Continuing on with the comparison, the next stage of AI we will witness will be that of early adulthood. At this stage, most AI tools will be more relaxed and will participate in society in a more beneficial way. Of course, this is my personal analogy of AI. As mentioned before, however, I don't expect AI scientists to build a perfect world of techno utopia. I also hope they won't destroy everything we hold dear and create a techno dystopia. We will probably find a general middle ground between these two extremes. Everyone's role is to ensure AI science does not result in a negative future, and to direct it to benefit humans for a more productive future. I believe we are at that moment right now. Our task is to clearly view this moment and participate in the works of the current generation to bring a better future.

Yoon: You're right. This is the moment. We must remember that today is the historic day for us to build a future of coexistence with AI.

Reich: I hope we can meet in person to discuss these topics. It will be wonderful. I would also like to talk to people in your world. You and anyone from any field are welcome to visit the campus I am at.

Yoon: Thank you for your time today in sharing your perspectives. I am grateful for the invitation as well. I will eagerly await the next time we get to speak to each other.

CHAPTER 3
Philosophers of the Era of AI

The World with the "Thinking Power" of Humans

We Must Solve Problems without Answers in Advance

AI (Philosophy) Framework
× Alison Simmons

Dr. Alison Simmons is a US philosopher who teaches philosophy at Harvard University. She is active in a variety of fields that have high interest in philosophy and human psychology with generation changes. She is the co-founder of Embedded EthiCS @ Harvard, which develops ethics modules for computer science curriculums and actively runs a variety of programs. She is also a faculty affiliate at the Department of History of Science at Harvard University. Her research focuses on modern and mental philosophy, which revolves around Descartes research. She has published many papers such as

"Cartesian Consciousness Reconsidered" (2012), "MindBody Union and the Limits of Cartesian Metaphysics" (2017), "Causation and Cognition in Descartes" (2020), and more.

Going beyond *Homo sapiens*, the Sole Humanity

Industry forecasts predict the emergence of "Robo sapiens," AI-based beings, coexisting with *Homo sapiens*. The coexistence of humans and AI robots is no longer science fiction. Although living with AI could add convenience to our lives, it could also lead to unimaginable chaos and ethical dilemmas. Questions regarding the distinctions between humans and machines, the nature of emotions, life, and other fundamental concepts will be challenged.

Consider a scenario where a brain-dead human with a functioning body receives an artificial brain transplant or vice versa. Which, if any, would be considered human? Such profound questions, including those about existence, responsibility, justice, life, and death, have long been debated in philosophy. However, these are not questions new to us in the age of AI. In fact, they have been pondered as philosophical topics since the beginning of civilization.

As technology advances, deep philosophical reasoning and ethical discussions are necessary to address these concerns. In this discussion, we will explore philosophical research in the context of AI and computer science advancements with Professor Alison Simmons of Harvard University. Drawing from Harvard's rich tradition in philosophy and humanities, she has developed the Embedded EthiCS program, a module-based course that embeds philosophical and ethical questions into computer science curriculums.

Together, we will examine the origins and evolution of Embedded EthiCS, which has pioneered ethics education in computer science. We will also discuss core topics related to the concept of being human and the relationship between humans and machines in the age of AI. Ultimately, we will explore the power of philosophy in providing insightful answers to the various issues we must consider in the future.

Harvard's Challenge: Embedded EthiCS

Yoon: Professor Simmons, I'm so glad you could join this discussion. As a leader in interdisciplinary efforts within computer science and AI, your insights, built over thirty years as a philosopher and professor, are invaluable. I am excited to discuss Harvard's Embedded EthiCS program with its founder and leader. Thank you for making the time.

There are many topics we could discuss, but I'd like to start with Embedded EthiCS. You and Professor Barbara J. Grosz,[23] both pivotal figures in AI, co-created this program. Embedded EthiCS @ Harvard is renowned for its high-quality lectures since its inception. I'd like to explore the program's beginnings, growth, and challenges and its current state.

Simmons: Thank you for inviting me to discuss this important topic. To talk about the origins of Embedded EthiCS, we have to go back to 2016. My colleague, Computer Science Professor Barbara J. Grosz, was teaching a course on "Intelligent Systems: Design and Ethical Challenges." The course attracted 144 students from various fields interested in AI and ethics, but only twenty-four were accepted due to its small seminar format. It was a group of students from computer science, philosophy, politics, and a variety of fields who had interests in AI and ethics.

During one class, Professor Grosz and her students debated the ethical issues surrounding Facebook's emotional contagion study and the general use of user profile data. Students became very upset with how user profile data was used and debated about its ethical issues. A few days later, Barbara gave them an assignment. "You are working at a social media platform company, and you have a new client who runs a fitness program. Based on your knowledge on social media platforms like Facebook, select and describe five user profile parameters to input into an algorithm for targeted advertisement." And the students passionately completed their work as they always did.

At the end of the assignment, Professor Grosz asked them, "How many of you considered the ethical meaning of the work you've done?" The result was zero. This realization alarmed Professor Grosz. She had high expectations for this group. So she called me and said, "Alison, I'm sending off students to Facebook and Google, but they don't know how to ethically think of the work they are doing. We need to solve this problem." Determined to address this issue before her retirement three years later, she reached out to me for help.

I liked having new challenges and felt that this was an important issue. I also liked Professor Grosz, so together we started our work. Together, we decided against creating a separate field of ethics just for computer science. Of course, those kinds of courses have their own purposes. But if we did the same, we were worried that we would repeat what happened with Professor Grosz's class.

We wanted to find a way to integrate ethical thinking into students' understanding of what computer scientists do. Then we came up with the idea of a pizza of embedded modules. We inserted modules into existing courses in relation to their contents. This way, students would be able to experience for themselves how issues related to their computer science studies happened. We thought we'd found the answer. However, we faced challenges, such as not being able to take over more than half of the other professors' courses. With only one or two classes available for our ethics modules, we needed a distributed model that would expose students to ethical reasoning across multiple classes. This way, they could repeatedly engage with ethical issues and develop a habit of considering them not only in AI and machine learning but in all areas of computer science, including hardware. That is how the Embedded EthiCS program came to be.

Yoon: I see. Thank you for the detailed walkthrough. I am reminded of how Harvard's Embedded EthiCS is an amazing collaborative work between two professors who answered the call to an urgent issue of our time. It looks like the situation Professor Grosz experienced in 2016 must have been something common in university computer science classes. The rapid pace of technological advancements in computer science for many years left little room for considering social impacts or ethical issues.

However, the emergence of AI has made it impossible to ignore these concerns any longer. In the past, students and scientists working in the field often lacked opportunities to question or fully engage with ethical considerations. Embedded EthiCS at Harvard is spearheading the change by teaching students to view ethical reasoning as an essential skill for computer scientists rather than learning ethics in isolation. The term "embedded" reflects the integration of ethics into the core curriculum, and this familiar term likely made it easier for computer science majors to embrace the concept. Could you tell us more about what happened after Embedded EthiCS program was applied to classes?

Simmons: The first time we started the Embedded EthiCS program was in the spring of 2017. We asked for help from four computer science professors and MIT graduate students who were interested in the program. Together, we developed four modules and opened four computer science major classes. We've grown a lot since then. So far, we've developed eighty-four modules and finished our thirty-seventh computer science major course. And now it has grown beyond our capabilities. We even had to turn away people because we simply could not take them all.

This is what happened so far at Harvard. Thanks to people like you, it's been growing outside of Harvard as well. Other universities are making use of our model. MIT, Stanford, Nebraska, Technion, and Toronto. We're also considering developing models fit for each specific university. They all have different situations. Each school has their own DNA, resources, and issues to solve.

So the challenge we have now is to help universities other than Harvard to develop their own versions of Embedded EthiCS. They will be their own programs and will work differently. For instance, how would you run the program with universities without philosophers? This is the kind of challenges we've discovered.

Yoon: The rapid growth of this field over a relatively short span of time is truly remarkable, highlighting its relevance in today's world. Harvard's pioneering efforts serve as a valuable model for other universities, fostering the growth and education of future generations. As a member of my field, I am eager to explore avenues for collaboration with other academic disciplines. Engaging in discussions like this one greatly aids me in discovering potential approaches and enhancing my understanding.

Philosophers and Engineers Live on Different Planets

Yoon: For our second topic, let's delve into your experiences and insights as a philosopher while leading the Embedded EthiCS program at Harvard from its inception. It appears that you and Professor Grosz had a strong rapport prior to collaborating on Embedded EthiCS. How did you find the experience of working with other computer scientists? Were there any challenges you encountered as a philosopher? I'm curious if there were instances where you felt as though you were speaking a different language or coming from entirely different perspectives.

Simmons: Interestingly enough, I only knew one computer scientist before that. Meeting others proved to be both exhilarating and eye-opening. We shared similarities, such as our problem-solving nature and preference for expressing our opinions. However, differences emerged as well.

Computer scientists typically seek definitive answers, whereas philosophers are accustomed to the absence of such clarity. That was the thing. Philosophers are used to having no definitive answers. This posed a challenge, but thankfully, most computer scientists were open-minded and receptive. Through Embedded EthiCS, we aim to help computer science professors understand that ethical considerations cannot be reduced to simple "programming rules" for an algorithm.

One notable strength of computer scientists is their innate curiosity. The unique advantage of Harvard lies in its identity as a hub for both humanities and sciences, attracting individuals from various disciplines, including engineers and scientists. Had it been a purely engineering-focused institution, our collaboration might have been significantly more challenging.

Yoon: It's great to see people accept their differences and work together. But I'm sure it's no easy task. Reflecting on my engineering background, computer science majors like myself were consistently encouraged to optimize practical functions. In contrast, philosophy grapples with abstract concepts such as right and wrong or life and death, which are inherently difficult to define. It's hard to envision how the thought processes and frameworks from philosophy could be directly coded.

In collaborating with computer scientists, what insights did you gain? Did you feel as though you held different worldviews or perspectives?

Simmons: Good question. Certainly, philosophers are not typically inclined to engage with coding. So we've come up with a method to negotiate our goals with computer scientists. One of my computer science colleagues described it pretty well. I can't remember the exact wording, but there were two important aspects. One was to secure conceptual resources to clearly express the issue, and the other was to know that there are people who are trained to think of those issues.

Our goal is not to compel computer scientists to code solutions for every problem. Instead, we aim to foster an awareness of potential issues and equip them with the intuition to recognize when something may be amiss. We provide tools that enable them to articulate their concerns and discern who should or should not be addressing particular matters. This way, they can communicate their insights, seek assistance, and engage with ethics committees.

The crux of this approach is empowering computer scientists to participate in these crucial conversations. As the saying goes, "Knowledge is power." By providing them with the conceptual resources to express and discuss their concerns, we hope to enable computer scientists to contribute meaningfully to these vital ethical deliberations.

Yoon: So the aim is for computer scientists and philosophers to share conceptual resources and collaboratively develop solutions by discussing and clearly defining issues together. That said, challenges related to computers and AI are not easily resolved. Engineers often seek definitive answers, preferring clear-cut solutions achieved through coding or eliminating bugs and errors. Facing situations without clear answers can be exceptionally difficult for them. Likewise, a program or machine that cannot determine the appropriate course of action in specific scenarios presents a significant challenge. Recognizing the difficulty of these tasks doesn't make the work any less arduous.

Simmons: Indeed, one approach that both engineers and philosophers can adopt is letting go of assumptions. For instance, while working on a project, we may have specific goals and concerns for certain users. We might take particular actions and explain why we coded something in a certain way. However, even after all this, issues may still arise. In response, we can reevaluate our initial assumptions, adjust them, and accept that mistakes will happen.

Or we could do it this way. This is an actual example from an Embedded EthiCS module. Let's say we're designing a video game. The game needs an inclusive design but has two different concepts for inclusion. How might these differing designs impact the final product? By exploring various assumptions, we can achieve better outcomes. In reality, we may have to choose just one design to sell as a product. However, by practicing these methods, we can first explain our choice, demonstrating thoughtful consideration. Second, we allow ourselves the flexibility to revisit and change that choice later on if needed.

Yoon: Your observation highlights the importance of remembering that society continually evolves. As a result, when designing programs, we should not only consider debugging tools for specific functions but also contemplate the ethical implications of our creations. Furthermore, if the initial assumptions underpinning a system change over time, we should be prepared to reevaluate and adjust them accordingly.

New Contemplations for Philosophers

Yoon: Many science and engineering majors might perceive philosophers as detached from everyday life, with their ideas seemingly far removed from practical concerns. However, as computer science and AI continue to advance rapidly, their social impacts can no longer be ignored. Consequently, computer scientists and engineers increasingly face problems that require philosophical and ethical considerations. This underscores the enduring relevance of philosophical values.

It's fascinating how many issues and dilemmas encountered in computer science and engineering, such as right and wrong, happiness, freedom, equality, and justice, have roots in philosophical inquiry. As for the future roles of philosophers, what do you envision their contributions will be in building a better society?

Simmons: I recall a student named Rachel who enrolled in my introductory philosophy course just for fun despite being a computer science major. One day, she asked to have lunch with me and shared her desire to study philosophy tailored for her generation—the tech generation. I encouraged her, acknowledging that such a philosophy didn't yet exist and that she would face numerous challenges. I emphasized that she would need to figure out how to overcome these obstacles. Rachel's perspective highlights the importance of developing a philosophy that is meaningfully applicable and closely related to our everyday lives.

Philosophy should be more accessible, as the work philosophers do is crucial. While some research remains within the realm of basic science and may not leave the laboratory, ideas can change the world. As you've mentioned, many fundamental concepts about ourselves are being challenged by today's technology, prompting questions about what it means to be human, the nature of life, and the essence of emotions.

These are precisely the questions that philosophers are trained to ponder. Therefore, I believe that philosophers should write as much as possible for the general public. While I may not be a gifted writer, some of my colleagues are. It's essential for philosophers to dedicate time to writing, even though we often find ourselves delving deeper into our areas of expertise. We must strive to break free from these depths. Fortunately, many of my colleagues have broad visions, and the younger generation is full of potential. This gives me hope for the future of philosophy. Now is the time where social dialogue is needed.

Simmons: During one of our discussions, you mentioned various devices that draw out emotional responses in people. In some cultures, evoking emotional reactions might be viewed positively, while in others, it may not be as well received and could even lead to feelings of shame. As previously mentioned, it's essential to recognize these differences during the development stage and adjust our assumptions to accommodate the specific context.

Yoon: That's not all. The issue becomes even more complex when we consider that determining what is appropriate or right can be difficult even within a single culture. Philosophers grapple with such questions, seeking rational explanations and answers. One example we discussed involved artificial beings designed to exhibit specific personalities and emotions. Whether machines can genuinely possess emotions or personalities is an entirely separate debate.

From a user's perspective, if such services or artificial beings demonstrate emotional empathy and users consequently trust or feel more confident in them, companies might capitalize on that trust to sell more of their services. Some may view this as exploitation, while others may see it as effective marketing. There is no clear-cut standard to differentiate between the two. If this trend continues, we may ultimately need a societal perspective to inform people about potential harms. In any case, it is a complicated question to answer.

Simmons: Indeed, this is an incredibly complex issue, which is why everyone must participate in the discussion, including those who write the source code for these robots and devices. It's crucial for them to learn and employ conceptual tools since they already understand what is practically possible and what isn't. Ideally, computer scientists, engineers, and philosophers should engage in dialogue alongside government representatives and the people who will actually use these technologies or devices. User participation is particularly important. Facilitating these conversations won't be easy, as technology is not confined to a specific field. It permeates our daily lives and influences everything we do. Therefore, we must foster communication at a societal level to address these concerns effectively.

Are We Really That Special?

Yoon: Now, I'd like to move on to the topic that you like the most. It's the field you've been researching for a long time: the philosophy of René Descartes.

Simmons: Sounds good.

Yoon: Although Descartes was a sixteenth-century philosopher, his arguments continue to pose meaningful and challenging questions in the twenty-first century, underscoring the importance and fascination of philosophy. He believed that the fundamental value of living organisms resides not in their physical existence but in their thinking souls.

Descartes equated human existence with the ability to think. In a sense, this idea relates to AI—a robot that can think. If so, should we regard AI as living organisms? Does this notion imply that AI could potentially embody superior morals and ethics, just as we expect from humans? How can we distinguish between humans and AI? And how should we differentiate ourselves from artificial beings? Can AI become human?

Simmons: These are important questions that can be viewed from various angles. But first, I'd like to point out something that's not quite like Descartes. It is something I find interesting and meaningful. In your questions, you described life and thought as the same things. One of Descartes's achievements was not only distinguishing the mind and body as separate entities but also completely separating life from thought. For him, they were fundamentally different things. Descartes was inspired by technology of his time, such as organs or fountains, to reach this idea. Life was purely a physical, mechanical phenomenon to him. Therefore, from Descartes's perspective, there was not much profound difference between living and nonliving things.

Unlike life, things like the mind and thoughts belong to a completely separate category. So, I think Descartes would've liked the idea of artificial lifeforms. That said, he would probably consider a machine that can perform life-sustaining functions, such as breathing, digesting, and reproducing, as artificial life. That would be what he saw as artificial life. But, as you know, he basically thought of organisms as just fancy and elaborate machines. He thought animals like dogs were living but not thinking. That's how he clearly differentiated life and thought.

On the one hand was biology, and on the other was psychology. So if we're to consider Descartes's influence, we must review again if we should think of life and thought on the same spectrum. In other words, in his perspective of life as mechanical beings, he wouldn't have any issues with artificial life. What I think would've intrigued him is whether the ability to think could be created artificially. He would've highly doubted the creation of an artificial thinking being, as no machine can think so flexibly or perform the process of learning, deriving concepts, and applying them.

Of course, Descartes could be wrong. We'll have to wait and see. We are not there yet. Descartes was also interested in using language meaningfully. Can AI understand and use metaphorical expressions or paradoxical expressions? This is a question Descartes would have raised on whether machines can really think. He would be happy to ask such questions.

However, there are current issues we need to be concerned about regarding the boundary between AI devices and humans. In the 1950s, the artificial respirator was invented, and today we are overcoming the COVID-19 pandemic because of it. Artificial respirators help patients with failing lungs and hearts to function normally. However, a question arose from this situation: Is the patient being kept alive by the artificial respirator, or is the patient actually dead, but their death is masked by the artificial respirator? It's a difficult question.

As a result, many conversations were held, and the very problematic concept of brain death emerged. Then brain death was introduced as a method to replace the declaration of death. The underlying idea was this: The brain operates the cardiopulmonary system. If the brain is dead, the respirator is just covering up death. But if the brain is functioning properly, it is an organ that helps the person stay alive. These two were distinguished as different things. There have been many debates on this thus far, and the medical definition of death has changed over generations.

Controversies related to brain death also raise these questions: What are we actually trying to save? Is it a person with conscious experience? Or is it just an organism with growing fingernails and toenails? It's a tricky issue. The example of brain death belongs to the field of bioengineering, but we will face these kinds of questions in many fields as technologies advance. Questions like "What is the borderline between what's living and what's not?" Many difficult questions without answers will follow, but we must continue to ask and ponder these questions together. They are unavoidable and most essential issues.

What could be more important than questioning the boundary between life and death? Doctors and families living at the forefront of this boundary have to face discomfort and confusion at every moment. When we face such core questions of our times, we should not hesitate or stop but continuously ask, research, and discuss with each other, even if it is difficult.

If it's OK with you, let's consider a slightly lighter example. I used to be interested in prosthetic arms. Prosthetic arms these days are made of the person's own biological material, which allows them to feel with that arm. Now, let's assume you have such a prosthetic arm. And you ate a cookie I had saved to eat. Then I got very angry and broke your prosthetic arm. You would also be angry and demand compensation.

Now, would you sue me for assault and battery or for property damage? Is the arm part of you or just a property of yours? When I point to my arm and say it's "my arm," we all know it's different from saying "my iPhone." Pointing to a part of my body is pointing to myself. But things can be different if it were a prosthetic arm. Therefore, we need to carefully consider various issues like these while developing technologies. Legal issues will continue to arise, and families will also need to decide whether to be sad or show different reactions in such situations. I feel that we want to solve this problem as much as we want to cure cancer. However, I believe that these issues should be considered not after they occur but from the research and development stage.

What Differentiates Humans from AI

Yoon: A famous phrase from one of Daniel Dennett's[24] books comes to mind when he says, "We expect people to behave in certain ways that we deem moral or ethical."[25] I believe the same expectations will be applied to those with prosthetic arms and artificial hearts as well. Despite having these artificial parts or organs, we will still consider them human.

However, if a person's nerve cells are replaced with silicon, should we apply the same standards to them? What about 40 percent or 51 percent of their nerve cells replaced? Should we view our brains, which are our "thinking machines," separate from the rest of our bodies? This leads into our next questions: If a robot or self-driving vehicle causes a crime, should we bring the device to court, or should we bring the engineer who developed the algorithm? Or should we bring the company or investor who created the device to court? This again returns us to the issue of discussing the nervous system, consciousness, and ethics. I think Descartes's distinction between the soul, body, and thought can be helpful in exploring this complex question.

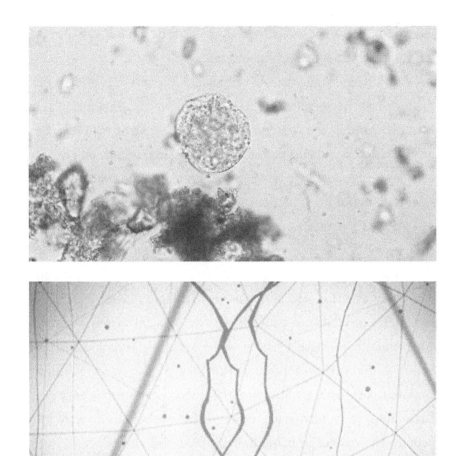

Descartes teases apart life and thought. Those are two fundamentally different things for him. Biology is on one side. Psychology is on the other.

Simmons: Yes, I agree. It would be great to have clear answers, but unfortunately, we don't. Two key questions emerge from our discussion: (1) What is our relation to AI devices? and (2) What are our responsibilities toward these devices?

While some argue that AI will replace humans, it is evident that there are tasks better suited for AI and others where humans excel. For example, AI lacks the common sense and goal-setting abilities based on free will that humans possess. Therefore, determining an effective collaborative model between humans and AI requires establishing a clear understanding of the roles and responsibilities of each party.

As AI continues to evolve and become more human-like, the complexity of this issue will only increase. Despite advancements, there will always be tasks that humans can do that AI cannot and vice versa. For instance, humans may struggle to find patterns in extremely large datasets, which AI can do with ease. Even if humans are eventually replaced by AI in some areas, significant differences between the two will persist. The idea of transferring human consciousness to an artificial brain and then into a robot presents a fascinating, yet controversial, concept. But we cannot do that with people. The essence of being human, including the ability to think subjectively and make decisions, may be difficult, if not impossible, for robots to replicate.

Our human form also plays a crucial role in defining our humanity. Intuitively, replacing all parts of a person with machine components seems to strip away their humanity. While minor changes, such as altering toenails, may not pose an issue, the deeper you dive in, the more complex the issue becomes. These are difficult questions to grapple with. Don't you think?

Yoon: Yes, I agree. The lack of consensus on what makes humans unique adds complexity to the issue. The debate over whether only humans possess consciousness or if animals, particularly mammals, also have consciousness remains unresolved. Moreover, various approaches to defining consciousness further complicate matters.

Some argue that much of human behavior and reactions are reflexive responses, not unlike coded machines. This leads to the question: What truly sets us apart? Identifying what makes us special and different is a challenging task. For example, consider the difference between AI-generated poetry and human-created poetry that stems from personal experiences. Can we confidently say that these two processes are fundamentally distinct? Without a clear understanding of what separates humans from AI, it becomes difficult to determine with certainty whether or not humans and AI are indeed different.

Simmons: I don't have good answers to these questions. But what you have said is indeed correct. We should be able to explain what differentiates us as humans. But it's not so simple. It is a task we must endlessly face. Descartes, for example, grappled with this challenge. His radical stance in his time was not due to his dualistic perspective, which separated mind and matter, or his belief in humans having a special mind. Instead, he was considered radical for asserting that most human reactions, including environmental and emotional responses, were automated like machines. Some behaviors may be educated actions, but he thought most were reactions to stimulus.

I sometimes ask myself: If Descartes lived in the current age, would he have been a neuroscientist or a computer scientist? I wonder how he would have answered these questions on problems we are yet to solve, such as those related to volition, reasoning ability, and flexible thinking system. What constitutes will and decision-making? While Codex,[26] an AI system, can write code, could it also set goals and rules through its own will?

This distinction might be one of the elements that make humans special. Despite appearances, we consistently hold one another accountable for our beliefs and actions. We might blame devices like Siri but don't hold machines responsible. Are we deceiving ourselves, and what is the root cause of this behavior? That's what we should find out. Examining our daily lives may be a good starting point. Reflecting on how we treat people differently from animals, such as dogs, or objects, such as phones, can shed light on our understanding of human uniqueness.[27]

Yoon: Those are profound questions. Like you said, we should start with the phenomena of our daily lives and dig down to their roots. Delving into their underlying principles can be a valuable approach, even if clear answers may not be immediately forthcoming. I believe that asking such questions and continuing to ponder them can have a fundamental impact on how we navigate through the complex situations we face and how engineers consider creating various tools and machines to be introduced to the world.

Simmons: Yes, of course. Engineers should be encouraged to consider, discuss, and develop technologies while asking, "How can advanced devices best serve humanity?" AI-driven devices are not necessarily designed to replace people, although some job displacement is inevitable. Ideally, AI will take over repetitive and standardized tasks, leaving humans to focus on more creative work.

We covered the value of work in one of our automated robot system modules for an Embedded EthiCS course. In this module, students explore crucial factors in differentiating between creative work and repetitive, standardized tasks, as well as the responsibilities of employers when promoting creative work and automating standardized tasks. They also consider the importance of a sense of systematic justice when deciding what to automate and what not to automate. Such discussions are not only interesting but also essential in informing real-world technological development and decision-making.

Yoon: I agree. It's true that we may be tempted to overlook or dismiss important considerations because they are too difficult or complicated. Since no single person can provide all the answers, we must engage in collective discussions to address these issues; otherwise, the responsibility evaded by our generation will impact the next.

As we live in the age of AI, people from all fields of expertise must explore and discuss numerous unanswered questions to harness the power of philosophy. By getting used to asking small questions in our daily lives, we can gradually tackle more complex and difficult inquiries over time.

Embedded EthiCS @ Harvard offers students a valuable opportunity to develop this questioning mindset. I eagerly anticipate the day when their experiences come to fruition, allowing them to address the uncertain questions of the future with strong philosophical foundations. This program at Harvard, along with similar programs elsewhere, holds great significance. I would like to express my gratitude to Professor Simmons for initiating and leading this impactful initiative.

Simmons: Thank you for recognizing and paying attention to our work. I am thrilled and grateful to have your support.

Are we really that special? If so, what makes us so special and different?

CHAPTER 4
The Beginning of Fusion

Engineering Is Great, but It Must Be Connected

AI (Engineering) Framework
× James Mickens

Dr. James Mickens is a US computer scientist and leader of the Embedded EthiCS program at Harvard University and teaches computer science at the Harvard John A. Paulson School of Engineering and Applied Sciences. He is also in charge of the Berkman Klein Center for Internet & Society. His primary focus is on the field of cybersecurity, and his recent co-authored papers include "Rethinking Isolation Mechanisms for Datacenter Multitenancy" (2020), "Identifying Valuable Pointers in Heap Data" (2021), and "Oblique: Accelerating Page Loads Using Symbolic Execution" (2021).

Acquiring a Broad Vision beyond Technology

We live in an age where human experience equals technological experience. Cutting-edge technologies, including AI, have become deeply integrated into our daily lives, enhancing convenience and dynamism. However, these advancements can also lead to unforeseen social issues. On a macro level, concerns arise regarding security issues between countries, potential terrorist acts involving AI devices, confirmation bias because of overreliance on algorithm-driven information, and the use of deep learning to create deepfakes that can fabricate scandals or mock minority groups.

As the AI era progresses, countries worldwide are working to establish legal and ethical guidelines for the proper and effective use of this technology. As a result of these efforts, the European Union (EU) Parliament announced the first AI legislative draft in 2021, containing comprehensive regulations for AI development and use. In regard to the EU's multinational response to AI-related issues, the US, Japan, and South Korea are also accelerating discussions on AI ethical rules that comprehensively ensure human rights and privacy protection, fairness, safety, responsibility, and sustainability of technology.

In this chapter, Professor James Mickens of Harvard University emphasizes the importance of integrating ethical considerations from the inception of technology creation and throughout the development process rather than solely addressing issues arising after implementation. Ethical considerations must be closely integrated from the moment ideas for technologies are born and throughout the entire development process as they grow. This approach highlights the need for computer scientists and engineers to develop ethical perspectives from their student years. Through our conversation with Professor Mickens, who leads the Harvard Embedded EthiCS program, we hope to inspire future leaders of the AI era to adopt a broader perspective, looking beyond the lab and considering the social impact of the technology they develop responsibly.

Attempts for Future Engineers

Yoon: Hello, Professor Mickens. We are delighted to have you join us today, and we appreciate your time. As AI advances rapidly, various philosophical, social, and ethical issues that we have never encountered before are emerging. It has become essential to share, compile, and discuss expert perspectives related to these concerns.

Today, we will explore various interests in ethics and AI. While other fields are crucial, computer science and engineering lie at the core of AI as they directly deal with and develop the technology. As both a computer scientist and an educator training computer science majors at Harvard, it is a pleasure and an exciting opportunity to learn about your thoughts and perspectives. First, let's discuss Embedded EthiCS, the collaborative program between Harvard's computer science and philosophy departments that you currently lead.

In the past, engineering majors had limited opportunities to encounter ethical and social issues, and approaching engineering with philosophical thinking and exploration is a novel and challenging endeavor. Could you introduce Embedded EthiCS from the perspective of a pioneer leading this change? I am curious about the nature of the program, how it is taught, and the students' reception of it.

Mickens: I am also excited to discuss these topics with you. Let's first examine the reason Embedded EthiCS was created. The primary motivation behind Embedded EthiCS was the growing need for engineers to be accountable for the social impact of their creations. This was particularly relevant to professionals in the computer industry, as our daily lives became increasingly intertwined with computer systems. Our current conversation serves as proof of this. Thanks to Zoom, we can comfortably communicate with each other from a distance. This app played a vital role in facilitating communication during the COVID-19 pandemic.

Additionally, consider situations involving making payments, using streaming services, and playing games. These interactions have become crucial in shaping how people connect with one another. In some criminal case sentencings, algorithms are used to determine parole eligibility, while banks use algorithms to assess credit ratings and loan amounts when processing loan requests. Consequently, human experiences are increasingly becoming computer experiences, whether we acknowledge it or not. That is why engineers, especially computer scientists, now have interesting questions they must ponder. These questions are not merely tied to technical aspects, but technology related to society. Ethical issues inevitably arise, posing high-level questions about what is right or wrong. While asking questions may be easy, the most significant problems often lie hidden in the finer details.

Building Environments to Think about What Is Right or Wrong

Mickens: The key objective of Embedded EthiCS is to create an environment that educates engineers and computer scientists to contemplate answers to questions like "What is right or wrong in this context?" "Who are the stakeholders?" and "Who should decide what is right?"

Moreover, we did not want to follow traditional engineering ethics models used in many universities. In general curriculums, engineers are required to take engineering ethics classes and review a few research cases. Many students view these classes as mere credit requirements, and they have little impact on them. After graduation, students often go on to write thousands of bug-free codes for companies without considering ethical aspects in their daily work.

While this approach is better than nothing, it does not teach students how ethical considerations should be embedded in their everyday tasks. We should not relegate ethical reflections to an afterthought, such as questioning the ethics of a completed design. Was my work ethical? Did I do the right thing for the customers, workers, and society? It is much more difficult to fix mistakes after realizing them.

At Embedded EthiCS, we ensure students do not feel "done" with ethics after taking just one class. We aim to embed ethical reasoning into numerous classes. Our fundamental idea is that if a computer science major student takes ten classes for their major, all of those classes should include some form of ethical reasoning. This way, even if the student does not take any ethics-focused classes, they will learn to regard what is right or wrong and who the stakeholders are as important perspectives.

This is the high-level approach we are attempting. We wanted to incorporate ethical components into various courses, be it computer interaction, compiler, or game classes. The ethical debates in our classes are based on specific issues related to the academic field. For instance, in a computer security class connected with an Embedded EthiCS module, we discuss the ethics of hacking a hacker.

Suppose a company falls victim to a cybercrime. If the victim knows who hacked the company's systems and tries to hack the hacker instead, is that appropriate? They could potentially retrieve stolen assets or forcibly shut down the hacker's operation. At first glance, this may seem like a simple question, but it sparks a fascinating and complex debate about cyberattacks and identifying the characteristics of hackers. When counter-hacking, how can you be certain the target is the actual culprit? What if you are wrong? What if the government tells you not to do it? Does that even matter? What role do legal regulations play in determining the rights and wrongs of counter-hacking?

Yoon: Indeed, that issue is quite complex with many related questions. Addressing these concerns with a fragmented perspective can lead to further complications. As computer science technologies are deeply intertwined with numerous aspects of our lives, it is essential for engineers who work with these technologies to be trained to thoroughly assess the impact of their decisions. Thank you for explaining the goals of Embedded EthiCS in achieving that. I'm sure it will be of great help for students reviewing ethical issues they may not have thought of.

Another Skill Essential to Engineers: Ethical Reasoning

Yoon: Ethics is an old field, with roots spanning thousands of years of human history. However, it hasn't always been the most popular. Now that examining ethical impacts has become important across various industries, we must consider how to increase awareness and encourage more people to understand and engage with ethics. As a professor teaching Embedded EthiCS to engineers who may be grappling with ethical issues they hadn't previously considered, what insights can you share about promoting a greater appreciation for ethics in these contexts?

Mickens: That is a good question. We aim to transform the mindset of students who may feel overwhelmed by ethical complexities. The primary objective of Embedded EthiCS is not to train engineers to write flawless essays on Nietzsche. Instead, our focus is on teaching them how to ask the right questions. In my opinion, one reason many find ethical logic and philosophy challenging is that most questions don't have clear answers. This can be particularly frustrating for engineers. I remember being in their shoes as a student. Back then, professors would give assignments like "Order this number sequence in ascending order."

Such tasks had clear answers, and we just needed to verify if they were arranged correctly. However, when faced with situations like an algorithm that prioritizes certain social media posts leading to negative emotions or conflicts, there's no clear solution. If someone claims that "most people think this video game lacks a representative character," is this truly the majority's opinion?

If we accept this assertion, what should we do? Should we view this as a zero-sum game or consider changing the character's design to make it more representative? These uncertainties can lead engineers to think, "Oh, I want to quit. I didn't come here to learn something like this." Through Embedded EthiCS, we encourage students to ask questions right away: Who are the stakeholders? How do we evaluate success? How do we define success?

By asking these questions, conversations can begin, and these conversations can have a significant impact. People in charge of ethical reasoning should play a role in groups, just as doctors and bioethics or philosophy majors are part of ethics commissions at hospitals. This prevents situations where engineers are forced to make decisions under challenging circumstances. Simultaneously, it's crucial for engineers to play essential roles in these conversations.

Of course, we don't want engineers to complete Embedded EthiCS programs and claim, "I know everything about ethics now. I don't need experts to help me." Instead, we want them to ask questions, learn basic terminology to understand and handle questions, and recognize when they need assistance from others in certain situations. If they can raise issues within the ethics commission or consider inviting experts on ethics, culture, diversity, or economy, they're heading in the right direction.

There is another point we emphasize to students in Embedded EthiCS. Recently, the media and politicians have been boosting the confidence of engineers. You've probably seen politicians say something in the line of "we need more STEM majors. Engineers are the future of our society." While this sentiment can be validating for computer scientists, it risks giving engineers the impression that technology can solve everything. I stress to my students that engineering is essential, but it doesn't exist in a vacuum. In other words, when considering the impact of the systems they build, engineers must listen carefully to ethicists or social scientists. This is because engineers create social technology systems that can significantly influence people's lives, and these influences extend beyond just technology.

Yoon: I agree with you. As a computer science major myself, I experience these challenges all the time. You've explained very well why ethical questions feel more difficult the more we think about them. The boundary between right and wrong can be difficult to define, and what may have been considered right in the past might no longer hold true today. You mentioned an example with video games, and I've encountered similar situations as well. I thought that ensuring equal representation of gender and race when creating game characters was the right thing to do.

However, I began to doubt my initial reasoning and wondered, "Why is it right or good?" It made me realize that my initial thoughts were not as clear as I had believed. Another complicating factor is that what was considered right thirty years ago may no longer be correct today. For instance, a decision made by the US Supreme Court in the 1960s could be deemed entirely wrong by today's ethical standards.

I believe that institutions like the Supreme Court need to adapt to changes in perspectives and standards so that they can review previous decisions and facilitate necessary social discussions. I would appreciate your opinion on this matter. From the standpoint that engineers should have independent perspectives, what do you think are the issues that we need to bear in mind socially or by government agencies in order to continue discussing these problems and social implications at this point in time? As a leader in the computer science field, please share what you would emphasize for society.

The Ultimate Goals of Embedded EthiCS

Mickens: That's a very important question. Let me summarize my answer into several points. One thing I think is important to note is that the purpose of philosophical and moral analysis is not to clearly discern right or wrong in a decision. In reality, it is rare for any engineer, or even any philosopher, to make such a clear judgment. This is an important topic we want to convey through Embedded EthiCS. In other words, if you show the same situation to several ethicists, it is highly unlikely that they will all make the same judgment about whether it is right or wrong.

Nevertheless, it is important to ask ethical questions because the idea of right or wrong is rooted in our values. You and I may also disagree on what's right or wrong in some actions, as we may not share the same values. One of the core points we want to convey to students through Embedded EthiCS is the ability for engineers to clearly articulate their own values. If they think, *we believe this is the right thing to do, but we're not sure why it's right*, they should be able to concretely explain why such a situation arose.

They should also understand what values they and others prioritize. This becomes particularly confusing when there are multiple stakeholders involved. If people with different values do not share those values when making decisions, friction arises in discussing what is right and wrong. In other words, it's important to understand that the goal of ethical reasoning is not dependent on something binary like "this is the only right thing" and to act accordingly.

This is where engineers, especially computer scientists, can become frustrated. We pursue clear answers and use binary logic, but unfortunately, that's not how the universe works. So, we must continuously think clearly about the values we prioritize and make sure they can serve as the basis for our decisions. Lastly, I want to say this: there might be an implicit view among engineers to avoid these social or ethical issues altogether by not taking any positions on them. I believe this is a completely wrong attitude.

Existentialist philosophers say that not making a decision is also a decision. For example, if you decide not to get involved in any social issue or event, it implicitly means that you support maintaining the status quo. There is a lesson from philosophy that I really like; it is a statement made by Søren Kierkegaard, an existentialist philosopher. He compared us to the captains of the ships. He said that we are the captains who have to steer the rudder, but if we decide to leave the rudder still, the wind will carry us to any shore it pleases. In other words, even the act of not deciding on a specific direction still moves us somewhere, and consequences will follow.

I sometimes hear engineers say, "I want to stay out of this debate. I just want to code." This statement reflects a sense of privilege, thinking that they can distance themselves from the social consequences caused by the devices or systems they design. I would like to tell engineers that they must remember their responsibilities and act accordingly.

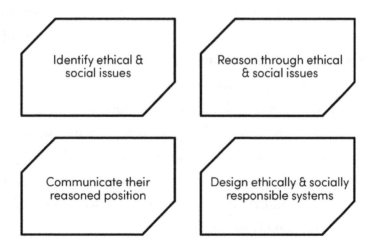

Goals of Embedded EthiCS Modules

Yoon: Thank you for your insight. I see that the points emphasized are well integrated into Embedded EthiCS programs. I understand that Harvard's Embedded EthiCS education and initiatives are not just confined to the campus but are expanding, so that more young engineers and computer scientists can become aware of its curriculum and make continuous efforts to understand and learn these crucial ethical approaches. As someone working on various aspects of Embedded EthiCS, could you provide some more details on this subject?

Mickens: Yes, we have formed a consortium with other schools that share our vision, providing educational ideas and sharing insights about what has been effective and what hasn't during the process. We offer several open-source courses, with modules integrated into various classes, such as databases, HCI,[28] or stand-alone ethics courses, all of which are available for anyone to access. This approach allows students to try different methods and find what works best for them.

Additionally, we are actively collaborating with academic fields beyond computer science. In the early stages, we worked with business and medical schools, as ethical issues are also prevalent in these areas. Through this collaboration, we found that technology can often serve as a factor that either mitigates or exacerbates risks arising from ethical concerns. I recall having numerous conversations with business schools in the beginning, discussing topics such as what technology start-ups should consider and how they should balance strategies to increase their user base or initial subscriptions with ethical concerns during user base expansion. Our goal is to facilitate communication on broad social impacts during the creation stage of technologies across all fields that discuss them.

Engineers Needed in Our Current Society: Coding Better Lives

Yoon: So far, we have discussed Harvard's Embedded EthiCS, a program that demonstrates your thoughts and passion. Now, let's expand our scope and consider the image and goals of engineers in today's world. I often encounter IT company CEOs, and their opinions on ethical decision-making in related matters can be quite diverse. Some argue that engineers should not be responsible for these decisions and that ethicists or others should handle them, while others believe engineers should concentrate on creating the most effective algorithms and optimizing technology.

I'm interested in hearing your opinion on this matter. Do you feel the industry environment has evolved as the applications of computer science and AI have broadened? How can we encourage society to view the social significance of engineering from wider perspectives?

Engineers Considering Ethics Issues from the Design Stage

Mickens: You've highlighted an important aspect we should consider collectively. Indeed, sometimes we encounter resistance from industry professionals and students who argue that philosophy is a separate field, and those interested should pursue it. But if coding is what you want to do, that's what you should be doing. Taking the field of medicine as an example, there's an interesting parallel. Do we prefer a doctor who only reads blood tests, or one who adopts a more comprehensive approach? Determining suitable treatments for a specific patient involves seeing the bigger picture, not just understanding the functions of various organs.

You need to see the bigger picture. Similarly, engineers must learn the fundamentals of their field, such as designing bridges for technical design engineers or understanding algorithm running times for computer scientists. However, if they don't address ethical issues during the design process and postpone them, they will risk creating systems that neglect important considerations, potentially leading to disasters. Afterward, they discuss, "What could we have done differently?"

In my opinion, engineers should spend more time reviewing and reflecting on the impact of the systems they build. While large corporations and small- to medium-sized enterprises should have experts or committees to address ethical issues, these conversations can also occur at the system design stage. As I mentioned earlier, it's easier to prevent disasters in early stages rather than after they occur.

For example, consider the spread of misinformation on social media. We are all aware of the severity of this issue and believe it needs to be addressed. However, if engineers had exercised moral imagination from the beginning when developing information delivery systems and infrastructure, they might have been able to prevent some of these problems. Instead of saying, "Just release it and let the market figure it out," or "The ethics commission will handle any issues," engineers should have been proactive in considering potential consequences.

Such attitudes are becoming increasingly difficult to justify, especially when creating products, operating systems, social media platforms, and games that millions or even billions of people may use. It is irresponsible to be excited about a product's potential impact while disregarding the ethical implications of the systems being developed. Consequently, many companies, particularly larger ones, have begun to establish internal ethics review committees or similar organizations to consider these perspectives. This shift is a positive change. Alongside these developments, we hope that programs like Embedded EthiCS can empower students and early-career engineers to contemplate and address broader ethical challenges from a more comprehensive perspective.

Engineers Taking Ethical Responsibilities

Yoon: I agree. But there are definitely concerns as well. Having an ethics committee with ethicists involved is, of course, a good thing. However, in social media, where various information and ideas are complexly structured and propagated, it's really difficult to predict what will have a consistently positive impact in the long term. The idea of having people with similar interests come together may seem like a pretty good idea, but who could have known that decades later, such a technical setup would result in echo chambers that excessively polarize political stances in the US election and make dialogue between them almost impossible?[29]

Similarly, as more and more people become exposed to their preferred and biased information due to algorithmic recommendations on the internet, they increasingly fail to distinguish between what is real and what is fake news or become unable to view situations in the broader context of the real world, exhibiting distorted perceptions. I think when an incident occurs, an ethics committee can make a decision. To argue that the decision is appropriate, it must be supported by quality information. However, as we've seen with what happens in echo chambers, if it is difficult to recognize or predict from the early stages the areas that may have significant ethical impacts, the value added by a separate ethics committee may be quite limited. In such situations, what can be done to create effective structures to provide appropriate safeguards for a company's technical efforts?

Mickens: Excellent question. I want to emphasize to future engineers engaged in this conversation that disasters can occur at any time, and the possibility is always present. As people commonly say, the systems we deal with often don't reveal their problems until something actually happens. I accept this as a premise because it is true. However, what this fact tells us is that, because of this, we should keep ethical considerations in mind more prominently at the designing stage. We should not be complacent even when a product has been approved and its ethical aspects have been examined during the design phase.

For example, think about continuous integration in the engineering field. What is continuous integration? It stems from the idea that an engineer can never have a truly finished product. Engineers regularly update and adjust their operation methods, continually monitoring the availability and performance of the engineering matrix. We should think about the social impact of technology in the same way. Despite our best efforts to identify stakeholders and potential negative impacts or social risks beforehand, we may still miss them in real situations. We are human, and we make mistakes; the important thing is to be open to correcting those mistakes.

Continuous integration is a widely used approach in software engineering today. It refers to the process of merging work copies from all developers into the shared mainline several times a day, ensuring that potential issues are identified and addressed promptly.

Using social media as an example, numerous data have already been discovered that could significantly change the way we operate our systems. You might have seen the *Wall Street Journal's* series of investigative articles called the "Facebook Files."[30] These articles revealed a vast amount of data accumulated within Facebook, such as information about how young women negatively perceive their appearances while browsing Instagram and the data on echo chambers collected by the platform. The articles also demonstrated how technical algorithm manipulation spread misinformation.[31]

Faced with such data, stakeholders might say, "We can't predict everything before the product is launched." That's a valid point, and engineers should accept that mistakes will always be made and may be unavoidable. However, this means that we need to continuously make inferences about what is right and wrong from a social perspective.

Another point I'd like to emphasize is that many people, especially those in companies or higher positions, are likely to say, "This system is just too complex. How can we possibly know what to decide?" However, the "Facebook Files" articles show us that there is a lot of measurable information available, and it's often clear how to interpret whether the impacts on us are positive or negative. It's important to note that it's not enough for engineers to simply engage in ethical reasoning.

Having ethical advisory committees is essential, but it's not sufficient. Above all else, what is crucial is having leadership that accepts these ethical considerations as a given. Without such leadership in an organization, most people will think, "There are these issues, but what can we do?" Even if others try to continue the effort, if higher-ups do not pay much attention, it will be difficult for change to occur.

Echo Chamber

The term "echo chamber" was coined by Cass Sunstein, a Harvard Law School professor who has primarily studied the "group polarization phenomenon" for over twenty years.[32] It refers to a metaphorical phenomenon where people with similar thoughts gather in a closed system, communicate only with one another, and strengthen their beliefs by repeatedly hearing similar ideas. As a result, they selectively accept and relay information that aligns with their positions. The likelihood of this occurring is higher in online media, especially in social networking services or online communities restricted to people with specific inclinations. With advancements in AI algorithms and the increasing provision of personalized information, more people are repeatedly exposed to content they prefer, making this issue more prominent in recent times.

By reinforcing cognitive bias, which only accepts and trusts information that suits one's taste and does not listen to opinions different from one's own, there is a risk of leading to extremism. The problem is amplified because it is almost impossible to verify or filter out fake news using AI technology when it spreads. The reason Professor Sunstein first used this term was to point out that the political confrontation between Al Gore and Bush supporters during the 2000 US presidential election had become excessively polarized, leading to division and conflict. Many also see the extreme confrontation between Hillary Clinton and Donald Trump supporters during the 2016 US presidential election as a result of the echo chamber effect.

The Facebook Files

In 2021, the *Wall Street Journal* published a series of investigative articles titled "The Facebook Files." These seventeen articles analyzed and summarized internal Facebook documents and included shocking information, such as Facebook intentionally leaving copyright-infringing content untouched for profit and neglecting the activities of drug cartels and human traffickers.[33] Among these, it is worth paying attention to the content related to AI technology.

Facebook has generally been known to use AI technology to remove harmful or unnecessary information from users' timelines. The intention was for the AI to review and block posts containing hate speech, discriminatory content, violence, or explicit materials. However, according to the *Wall Street Journal*, Facebook's AI was almost unable to distinguish between normal posts and problematic ones. According to an internal Facebook report, Facebook's own AI was able to remove only about 2 percent of posts with hate speech that violated its rules, while the rest were addressed by manual review, for which the company reportedly spent around $2 million per week.[34]

Furthermore, it was revealed that Facebook designed its algorithm to prioritize posts with more negative reactions by assigning 1 point to likes and 5 points to dislikes, causing posts with negative reactions to appear higher in users' timelines. In addition, Facebook managed about 5.8 million celebrities separately through a program called "Cross Check" or "X-Check," granting them privileges and exempting them from strict content moderation rules.[35]

It was also reported that the company suppressed internal research findings on the negative impact of Instagram use on the mental health of teenagers. Of course, Facebook immediately issued a statement refuting these claims as false, but the controversies surrounding online systems, AI, and advanced technology are not easy to resolve, and the debates sparked by "The Facebook Files" continue because of the ambiguous boundaries between right and wrong.

Engineers Optimizing the New Age

Yoon: I completely agree with your statement that integrating an ethical perspective sufficiently in the engineering process depends on leadership with such a vision. However, I also believe that it's important not only for engineers and ethics committees, but also for all of us involved in the process to pay more attention to ethical issues and develop habits of thinking about social impacts.

For computer scientists and engineering majors, the most fundamental aspect of engineering is optimization. Engineers like us have always been trained to optimize systems for specific purposes. This can increase execution speed and system efficiency. The ability to quantify and understand these improvements is also an advantage. However, if we now need to consider other variables, including ethical reasoning, systems or algorithms that are not optimized for traditional variables might emerge. Could this mean that engineers need to be open to embracing various methods? I think this direction is a significant departure from traditional engineering education, where the focus was on objective optimization.

Mickens: You're absolutely right. In some ways, for engineers from the old era, this approach might feel like a complete rearrangement of their engineering concept. It is indeed a big change, and there may be a lot of resistance. Most engineers are concerned because it's difficult to evaluate what reasonable reactions are. They struggle with abstractly important but ambiguously unquantifiable aspects like diversity, which don't fit neatly into engineering terms.

However, the interesting part for me is that when you look at a company making a lot of money from targeted advertisements, you won't hear such statements from anyone in that company. Statements like "Gosh, ad targeting is so complicated. We give up. We really can't do anything about it." They'll somehow find a way to accept ambiguous concepts. Starting with measurable aspects such as ad click rates, they'll go beyond just focusing on the numbers and ask whether those figures truly represent the best ads for people. Then they will meticulously question whether the ad is really the best or how to define what is best. In fact, these large tech companies pour literally billions of dollars into designing and operating such massive programs.

Another thing we should clearly acknowledge is that tech companies, large or small, generally don't shy away from problems when they arise. That's the remarkable aspect of tech companies. And it's also why I'm passionate about technology. Engineers like us dream of providing email services like Gmail in a more scalable way or building internet infrastructure that can deliver hundreds of thousands of page views on a website within seconds.

Of course, tasks like these are incredibly challenging and difficult. However, as an engineer, I enjoy tackling them head-on. We should say, "This is a difficult problem, but it will work out, and we will eventually find a solution." Developing technology that is beneficial to society is certainly not an easy task, but I believe there is no reason to avoid it just because it is difficult to quantify.

For example, think about the infrastructure of real-time ad delivery networks and machine learning algorithms for natural language translation. Web browsers, operating systems, and games built by modern engineering are usually composed of hundreds of thousands or even billions of lines of code. These systems are well deployed, allowing us to rely on things like internet routing infrastructure today. It is undoubtedly a complex system. An engineer once told me this: "I don't know how to optimize for these social issues because they're so ambiguous. I want to do something simpler." But that's not possible, because modern society is complex.

I sometimes hear opposing opinions from old-school engineers about my perspectives, saying, "In the past, we could focus only on the amount of information being processed, and that was enough. It was much simpler back then because we didn't have this kind of technology." However, just as doctors from the old days used simple folk remedies, but in today's world with a deeper understanding of the human body and medical techniques, they have evolved to adjust and utilize various treatment methods including folk remedies, I believe the situation is similar for engineers. The society we live in today is fundamentally different from thirty years ago.

I am part of the amazing generation that experienced that change. When I was young, not every household had a computer. Around the time I was five or six, personal computers like the Apple II started to appear. I experienced these changes, and now we live in a completely different world. Therefore, we must approach engineering in a different way as well.

Yoon: Indeed, it is not easy to develop better algorithms in the face of complex and challenging situations, particularly when the rewards might not be immediate. Rewards could come in thirty years or even centuries later, much like the efforts to combat climate change. It is difficult to unite people's efforts and attention toward these issues, as opposed to traditional problems that offer immediate rewards. However, if more people and young engineers continue to pursue their passion for more sustainable and superior algorithms despite these barriers, and if they come together, dedicate time, and persist in making genuine efforts, I believe that it will undoubtedly bring positive effects to society.

Mickens: Discussing climate change truly hits close to home. In my opinion, issues that our generation once considered concerns for ten to thirty years in the future have become immediate problems for the younger generation and engineers today. Climate change is undeniably an urgent issue, as we are already witnessing the effects of global warming. Even if these events unfold over a longer time and on a larger scale, we are already on that timeline. Moreover, issues related to online harassment and the spread of false information are also part of our current reality.

From a shortsighted perspective, companies might not see addressing these issues as urgent. Focusing solely on maximizing shareholder value may benefit the company in the short term, but in the long run, it will bring more harm than good to both the company and society. A change in attitude is needed. The encouraging aspect is that not only young engineers but also senior engineers with extensive experience in the field are creating a ripple effect. It seems they, too, have realized that we are now living in a new world.

Thirty to forty years ago, mainframe computers were used by stock traders and researchers simulating nuclear explosions. Engineers of that era might have thought their coding designs wouldn't significantly impact society. However, that perspective was only valid because it was thirty to forty years ago. Not anymore. Today, we carry around devices in our pockets that would have been considered supercomputers just fifteen years ago. It's fascinating but also concerning. Consequently, engineers who believe their work has no social significance are mistaken even from an experiential standpoint. That's why we're striving to change this passive mindset through Embedded EthiCS and other various activities, and we will continue these efforts in the future as well.

Yoon: Indeed, it is a vital matter for both the present and future generations. Having this conversation with you today has given me a more hopeful perspective on the future of engineering. You are putting in significant effort to promote the values you stand for, building a team of embedded ethics leaders who share your vision for a better future, and fostering the growth of young engineers. As a businessperson, I also hope to maintain the same goal and continue to provide and receive the necessary support to collaboratively address these ambiguous and complex issues.

Mickens: It was great talking to you. Thank you for inviting me.

Yoon: I hope we can continue our discussion sometime in the future. Thank you for lending us your time.

Drawing the Line in a World without a Boundary between Right or Wrong

CHAPTER 5
Unheard of Questions

*As Long as We Talk to Each
Other, There Is a Chance*

AI (Society) Framework × Alex Byrne

Dr. Alex Byrne is a US philosopher who has served as the head of the Department of Linguistics and Philosophy at the Massachusetts Institute of Technology (MIT) and is currently a professor of philosophy. He has spent a significant amount of time exploring the fields of linguistics and philosophy, and is actively engaged in research and writing on self-awareness and epistemology, based on the philosophy of mind. His books include *Transparency and Self-Knowledge* and *Perception and Probability*, and his recently published papers include "Perception and Ordinary Objects" (2019) and "Concepts, Belief, and Perception" (2020).

The Age of AI without Right or Wrong—the Role of Philosophy

Humans are filled with numerous biases and contradictions. This also means that AI, programmed to mimic and learn from humans, can amplify and reproduce human biases. However, the subject of human bias and prejudice involves many intertwined questions without clear-cut answers. Of course, if a certain bias or prejudice poses a risk of unfairly treating a specific group or individual, it is right to approach it with caution and address its negativity. Still, the question of whether bias or prejudice itself is good or bad is a separate issue.

If bias is not necessarily bad and may be something humans possess out of necessity, should it be directly reflected in AI as well? Or should prejudice be completely removed when applying it to AI, given its mostly negative aspects? Can we technically control the element of bias? Even if we can, is manipulating it the right thing to do? Can we call an AI free of prejudice a more fair, ethical, and good AI?

In the technologically driven society that AI will lead, numerous philosophical and ethical issues do not have clear distinctions between right and wrong. At this point, research on technology ethics is emerging as an important issue, beyond merely focusing on technological innovation. Modern technology, including AI, has deeply permeated daily life to the extent that it impacts the lives of the entire human society, not just staying in the realm of functional improvement. In this conversation, we will explore various topics related to AI and technology ethics with renowned American philosopher and MIT Linguistics and Philosophy Professor

Alex Byrne. Through our dialogue with Professor Byrne, we hope to examine multiple perspectives on various questions about AI technology and societal ethics, such as what ethical standards should be applied to AI, whether completely removing prejudice is the solution to creating an ethical AI, and how we should address the diverse ethical dilemmas surrounding AI and the potential impacts on our society.

What Is the Right Choice?

Yoon: Professor Byrne, I'd like to thank you for making the time to talk about AI technology and ethics. I would like to explore with you various ethical dilemmas and potential impacts that may arise from AI technology and discuss what kind of considerations we should make and what alternatives we can create within our societal structures. Above all, it is an honor to have a conversation on this topic with someone as distinguished as yourself. Before we begin our conversation, could you first introduce your research interests?

Byrne: I'm glad to have the opportunity to talk with you. Currently, I am a philosophy professor at MIT and have previously served as the department head for MIT's linguistics and philosophy program. It is quite unique to have two graduate programs combined and operating with separate faculties, as we have done. My research primarily focuses on the philosophy of mind, examining how human psychology connects to the real world and the brain. I have also conducted research on self-awareness, a theory related to how humans perceive their own mental states. I have authored books on this subject as well. Recently, my interests have expanded to epistemology, which deals with the question "How do we know what we know?" While

I may not be a traditional expert in technology ethics based on my career alone, the field of technology ethics is still emerging and lacks many established experts. Therefore, I believe that I can contribute meaningfully to the conversation on this topic.

Yoon: Of course. I'm sure your experience and research will be more than enough to provide meaningful insight on this topic. From the perspective of someone who has long studied philosophy and linguistics, what are your thoughts on the recent development and proliferation of the AI field related to computer-human interaction and technological advancements?

Trolley Dilemma

Byrne: It's true that the AI field has made incredible progress over time, and this has had a profound impact on our lives. The development of AI has brought about significant advancements in the field of robotics, and it is now impossible to implement robotics technology without AI.

One area related to AI that philosophers, in particular, find intriguing is self-driving vehicle technology. This is because it is directly connected to the famous "trolley problem" proposed by MIT Professor Judith Thomson.[36] The "trolley problem" is essentially as follows: A trolley is hurtling down the tracks at high speed, with five people tied to one side of the track and one person tied to the other side. If nothing is done, all five people will be struck by the trolley and die, but if you pull the switch to change the direction of the trolley, only one person will die. The question is whether it is morally permissible to pull the switch and, if so, the ethical reasoning behind it.

In reality, there are numerous complex and diverse variations of the trolley problem. As you mentioned, this is also related to the software programming of self-driving vehicles. It involves dilemmas like how to program self-driving vehicles to react in certain situations, such as when changing direction could result in the death of one person on a bicycle while saving five people on the sidewalk.

Yoon: The trolley problem is indeed a fascinating thought experiment that prompts us to consider important ethical questions. People often struggle to make decisions in such critical moments. I recently read a book discussing how to differentiate between right and wrong. What I believe to be right may not align with my neighbor's views, and similarly, even my family or friends may not share the same values if they live on a different continent or in a different climate. Ultimately, we cannot be certain of what is right or wrong, and we cannot always have confidence that our judgments are objective.

Given that humans themselves are flawed, we cannot assume that decisions made by AI, which are modeled after human intelligence, are always correct. So, is it possible to design AI in an ethically sound and socially acceptable direction? What steps should we take, and how can we ensure that AI is programmed in such a way that it aligns with these ethical and societal expectations?

A Trolley is running at high speed. Five people will die if it continues on its tracks,
and one person will die if you change the direction. Will you pull the switch and
change the track or not? What is the right choice?

Byrne: Great question. In fact, such questions have existed for
a long time. If you're familiar with Isaac Asimov's science fiction
novels, you'll recall the Three Laws of Robotics that Asimov proposed.
Although I don't remember the exact laws, the first one is something
like "A robot must not injure a human being." This principle has
become a fundamental rule for suggesting how robots or AI should
behave, and many papers on the subject are still being published to
this day. Of course, no definitive answer has been given as to what the
most correct and best solution is.

In general, it is not easy to reach a consensus on the right policy
in any society, especially in a democratic society. The same would be
true even if the scope is narrowed down to specific ethical issues. For

example, "What is the right decision in this situation?" "What policies should we adopt from an ethical perspective?" or "What does justice require of us?" are such questions.

Is it just to take money from the rich and give it to the poor? If the wealthy earned their money honestly and did not steal from others, then taking their money without permission would not be a just act. These kinds of issues are complex problems that political philosophers have been discussing for centuries. I don't see AI as raising new kinds of problems in this regard. Ultimately, at some point, we will have to accept each other's differences or reach an untidy compromise about the direction we should move forward.

In the movie *Saving Private Ryan*, soldiers are deployed to save just one private, and from the general's point of view, such a decision was quite justifiable both strategically and symbolically. However, if we only count the number of people who were sacrificed to complete the mission, it is still questionable whether it was the right decision to make. Similarly, the ambiguity of right and wrong you just mentioned is very closely related to the trolley problem.

It demonstrates how difficult these kinds of problems can be. Let's assume we ask people, "If you were in a situation like the trolley problem, would you pull the switch to save five people and sacrifice one?" Most people would say, "Yes, I would pull the switch. Although one person would die, at least it wouldn't be the wrong choice." Choosing the option where fewer people are sacrificed can be considered a better principle. So, you would follow that principle. While sacrificing the minority over the majority can be one principle, further consideration may reveal that it may not always be the best principle.

There is a related example that you might have heard of. Let's suppose a healthy man visits the hospital to see his mother. There are five patients in this hospital. They each need an organ, such as a heart, liver, or kidney, and will die soon if they do not receive a transplant. If we were to divide the organs of this innocent man who came to see his mother among the five patients, we could save five people by sacrificing one person. However, when asked whether it's right to kill an innocent person to save five others, most people would say it's definitely not.

Therefore, the choice to kill the minority to save more lives is not an absolute principle. Principles are much more complex. That's why ethics is difficult.

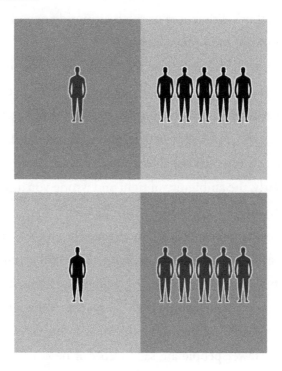

Can we kill an innocent person to save five dying people? The principle of saving many over a few is not an absolute principle.

Dilemma of the Age of AI: The Starting Line to Solve Problems

Yoon: Indeed. These ethical issues are not new but rather have always been present around us. However, as large tech companies have begun to apply specific ethical standards or decision-making processes to software or services that significantly impact daily life, people generally expect these companies to meet certain criteria. The difficulty today lies in the fact that it is unclear what these criteria are and where they originate from.

The spread of echo chambers, where people selectively accept only the information they want, and the extreme division of society as a result of individuals being trapped in their own perspectives and cutting off conversation seem to be related to the widespread use of certain criteria without social consensus due to technological growth. Therefore, I believe that some form of social agreement is urgently needed to minimize the potential negative impact that may arise from the adoption and application of advanced technologies.

Byrne: I agree. However, the existence of a consensus is not always at the core of issues. A consensus may already exist regarding acts that everyone agrees are harmful. The reason why negative situations persist is that there is no incentive structure in place to prevent them.

Let me give you an example. Recently, it has been reported that Facebook applied a specific AI to its Instagram app, further confirming its usage.[37] This AI has the ability to display various images based on the user's past browsing history and interests. There is evidence that such features are harmful to the mental health of teenagers, especially teenage girls; it can lead to issues related to body image, as children become overly concerned with their appearance and the evaluations of others.[38] Everyone agrees that this is bad. There is complete consensus on that aspect.

The issue is, how do we stop this from happening? This actually happened, and Facebook knew very well this was a negative thing. They had an internal research report stating that this algorithm is harmful to the mental health of young girls, but there was no effective incentive in place to stop Facebook from using this function. So, I think that for many of the issues surrounding AI today, rather than focusing on getting people to agree that it's bad or harmful, it's more important to establish a structure that can prevent such harm before they happen.

Yoon: You mean to change the direction of incentives?

Byrne: Correct.

Yoon: I agree with what you've said, but the majority of companies, especially CEOs and executives, have internal regulations that they are responsible for implementing. There are clear regulations that punish them if they don't act in the direction of maximizing shareholder interests, and if they violate their duties as corporate fiduciaries, they actually face penalties. However, there are no legal or policy sanctions to regulate or punish corporate executives that don't prioritize the mental health of teenage girls.

Do you actually see any movements in our society to review policies and change the incentive structure in the direction you've mentioned?

Humans and AI: Work to Coexistence

Byrne: We can say that our efforts in teaching technology ethics at MIT are part of such movements. Many MIT students will become software developers and join companies that create platforms like

Instagram. The core of MIT's technology ethics education is to help students who join companies like Facebook to think thoroughly about the consequences of their work and develop a sense of ethical responsibility for the outcomes of the technologies they develop. We are confident that this will be helpful.

However, as you mentioned in the case of corporate regulatory laws, technology ethics education at universities cannot fully solve the issues we are considering. This is because there are social dilemmas and collective action problems.

It is not enough for every individual to have good intentions or to not intend to harm others. In reality, there are occasionally flawed incentive structures. So it is impossible to collectively form policies that genuinely help everyone. Ultimately, the educational challenge when teaching technology ethics goes beyond merely exposing individuals to ethical perspectives or helping them understand the relationship between ethics and technology. Perhaps technology ethics education is about cultivating citizens who realize that being a good citizen in a democratic society involves engaging in collective actions to establish necessary regulations. In some situations, the free market economy may be able to solve problems. However, regulations are needed even for the free market economy to function properly. Ultimately, this is not just an individual issue but a societal issue as a whole.

Yoon: But since society is made up of individuals, we definitely need educational efforts change their perspectives.

Byrne: Yes, you are absolutely right. What I want to say, however, is that we cannot solve the problems of our society just by teaching individual students ethical perspectives.

Yoon: I see. The issue is a lot more complex and difficult than I expected. Across society as a whole, it's important to emphasize the need for ethical perspectives, especially in technological fields centered around AI, which involves considering the social impacts of these technologies and forming related policies. It is also crucial to establish legal regulations that ensure appropriate responsibility to their outcomes.

Is AI without Prejudice Better?

Yoon: Even for seemingly simple concepts like fairness, what is considered fair in one group might not be perceived as fair in another group. From which perspective can we say that fairness is most important? And who can decide that? I would like to hear your opinion on how we should approach such issues.

Byrne: There is an extreme form of thought that claims objective right and wrong do not exist at all. This perspective sees everything as relative. What may be right according to my values might not be right when viewed from someone else's values. And there is no neutral decision-maker who can draw a conclusion.

They believe that there is no one who can decide that one person's values are right and another person's values are wrong. If you think from such an extreme relativist standpoint, you cannot express any opinion. It becomes meaningless. And it just turns into a massive power struggle. It becomes a matter of who will win by force or who will win by debate, and in the end, it becomes a matter of who will win.

Of course, even after we agree to discover something through discussion and debate, there will undoubtedly be no easy solution. You can't just pull a book off a shelf and say, "All the answers to such ethical problems are here." However, at least by sharing conversations and engaging in debates together, there is definitely a possibility of discovering something. I would like to look at it from that perspective.

Can We Demand Ethics from AI?

Yoon: A key issue that often arises when discussing AI solutions is that AI operates based on human data, which inherently contains various biases. As a result, AI may amplify and reproduce these biases. Given that humans are not free from biases, can we really expect higher ethical standards from AI?

Byrne: You've raised a valid point. Many people have indeed published articles expressing concerns related to AI, including those related to facial recognition programs and AI that review résumés of job applicants.

It might be helpful to differentiate between technical and philosophical issues in this context. A technical issue would be when a facial recognition software fails to accurately recognize certain races or genders. In this case, the problem is not the software's purpose but rather its technical limitations. No one would want facial recognition software that cannot accurately recognize specific groups, so the issue would be improving the software technically.

However, when we operationalize the concept of fairness differently, it becomes a philosophical issue. There are various mathematical methods to operationalize the concept of a fair outcome, and these methods are not all compatible. Choosing one means giving up another. If you prefer one approach, you have to give up another way of operationalizing fairness and make a decision at that time. This case can be considered an ethical or philosophical issue rather than purely technical.

Many issues related to facial recognition software, like the MIT Gender Shades Project,[39] are mostly related to technical problems. These issues are relatively easier to address compared to ethical issues. Of course, solving technical problems can also be very difficult, but there is a certain agreement among people who do not object to the social consensus you mentioned earlier. However, there are much more challenging issues where even such social consensus does not exist.

Yoon: I agree that engineers and tech company executives may not always prioritize ethical or philosophical standards during product development. As you know, they often focus mainly on optimizing technical outcomes. However, when we create machines that make specific decisions in specific ways within certain environments, it becomes crucial to equip them with criteria for making decisions in unforeseen environments or situations.

In such cases, the question arises whether robots or AI should possess higher value standards than human ethical standards or behave like the majority of humans, with all sorts of shortcomings and biases. This issue eventually leads back to the question "What is a better outcome for AI?" Until now, we have measured AI's performance with relatively linear criteria. If we expect a higher level of ethical standards from AI, should the way we evaluate AI also change?

Byrne: There are indeed cases where AI can surpass human performance. While AI may not be able to outperform general human abilities, it could excel in ethical decision-making compared to human judgment. Let me give you an example. If I have a bias when evaluating résumés of women and men, we could theoretically program an AI to follow the same decision-making process as me but without the gender bias. In this case, the AI would become a slightly better version of myself but without a preference for men. This approach could be one way for AI to achieve better outcomes than humans.

Of course, AI has the potential to think qualitatively differently than humans when it comes to ethical reasoning. It could make choices that we might intuitively consider wrong, yet it may have a deeper understanding of ethics, the way ethical rules work, and how they maintain balance with each other than humans do. It is similar to how computers can perform mathematical calculations impossible for humans to accomplish.

Humans are amazed when computers find the answer to complex calculations, leading us to become fascinated with the idea that AI can excel in the ethical domain as well. We may simply admire the solutions AI comes up with and think we should proceed with them. Who could have imagined this? I never thought such demands would emerge in social and moral aspects. This seems to contradict human intuition. However, now we must trust the choices of machines. While this idea might be just a fantasy, it's worth considering whether such a situation could genuinely happen from a theoretical standpoint. It involves the assumption that the fundamental truth of ethics is too complex for ordinary humans to grasp, and only a powerful AI can truly understand it.

AI and Prejudice

Yoon: It's true that humans have limitations and biases, which are often reflections of our evolution and experiences. These biases and shortcomings might also be shortcuts we've acquired from living in specific environments for years. To pose a question for the sake of questioning, how can we be sure that the complete absence of bias is a better strategy or the right thing?

What if, for some reason, our society evolved in a direction where it carries more biases? If we cannot know what is truly right or wrong in an absolute sense, then everything becomes relative. And if everything is relative, how can we say that the complete absence of bias is always right, better, and a more superior ethical value? Therefore, I believe we find ourselves in yet another dilemma.

Byrne: You're right. The issue lies in the meaning of bias itself. Bias is not inherently bad in every situation. For example, if you have a bias toward good grades or impressive work experience when reviewing résumés, that's a good thing. However, if you prefer résumés of a specific gender or race, that's problematic.

So, if we understand bias in a neutral sense, good judgment involves some level of bias, and that's nothing to be ashamed of. Generally, bias means taking into account factors that are irrelevant to the decision at hand. If we can discern what these irrelevant factors are, AI itself would be well suited to address bias issues by excluding these unrelated factors. This contrasts with ordinary people who often struggle to ignore factors unrelated to their judgment, even when pointed out.

Efforts toward Ethical AI

PREPARING FOR AN ALLIANCE WITH AI

Yoon: We've now come to the last topic we'll discuss together. As we face the inevitable future of the AI revolution, what preparations should we make for ethical AI technology, and what roles should philosophy and philosophers play in this process? How do you view the current problems our society faces from a philosopher's perspective? And how do you think these problems differ from those faced by humanity in the past? I'm also curious about your position. Are you optimistic that we can solve these problems?

Byrne: That's quite a broad range of questions. First of all, I am not qualified enough to answer questions about what will happen in the future. When I first came to MIT in the 1990s, my office was using an early form of the internet. At that time, I told my colleagues that the internet would never become widely popular. So, I am not particularly good at predicting the future.

Looking back at the terrible performance of many people in predicting future trends in society and politics, it is evident that predicting those trends is much more difficult than making predictions about climate or the spread of diseases like COVID-19. Just with COVID-19 predictions, many pathological models turned out to be completely wrong.

One of the reasons why predicting the future is especially difficult is due to the dramatic discoveries or advancements in the AI field that we may not foresee in the near future. Such advancements do not naturally occur from what we already have. Someone might make a completely new and significant discovery, or a new technology like the invention of the transistor. Human communication is a complex, interrelated system, so predicting it is incredibly difficult.

Additionally, the factors I mentioned earlier will only make future predictions even more complex and challenging. Though I cannot be certain about what the future holds, I don't think we need to be pessimistic. I prefer to see the glass as half full rather than half empty. AI technology undoubtedly has tremendous potential to positively impact society.

Yoon: I see. I appreciate your answer.

Byrne: I actually wanted to ask your thoughts. How would you answer your own questions?

Yoon: As you mentioned, I too am uncertain about what the future holds. However, it is crucial to acknowledge that the social impact and ripple effects of AI technology have grown substantially compared to the situation twenty to thirty years ago and the predictions made back then. We all need an appropriate framework to discuss and carefully consider the ethical implications and social impacts of technology. This framework is an essential tool for engineers and business managers, allowing them to think about and understand societal impacts before implementing any algorithm in practice. This is why I believe it is even more important for undergraduate students to learn about these tools before entering a complex and uncertain society. Equipping them with this knowledge will enable students to ask questions and approach problem-solving with a more responsible attitude, ultimately helping to shape a better future.

Byrne: It gives me strength to hear that. It is very important for people like yourself to keep voicing these issues.

LOOKING AT THE AGE OF AI THROUGH PHILOSOPHY

Yoon: It seems that we have delved into questions that do not have clear answers today, such as distinguishing right from wrong and determining what is biased or fair. At the same time, we expect machines and algorithms to possess clear judgment on these open-ended issues. Does this mean that the traditional approach to developing new technologies and advancing society with those technologies needs to change? What are the social discourses for reaching a societal consensus and increasing understanding about the future we will face, and what do you see as the role of philosophers and ethicists in this context?

Byrne: Your question touches upon two issues: one regarding the role of philosophers, and the other concerning the extent to which we delegate decision-making to AI or supercomputers. As you know, most AI is utilized as a tool for identifying various patterns in data or sorting data along different dimensions. Ultimately, it is humans who make meaningful decisions based on the information extracted through AI. So, an abstract question to consider is "In what situations will we allow machines to make decisions?"

For instance, imagine you have been sentenced to six months in prison. Based on past criminal records, a machine predicts the possibility of recidivism. Humans don't necessarily follow the machine's opinions or suggestions but instead make their own decisions based on the results. This represents one of the issues at hand.

Another entirely different issue pertains to the role of philosophers in all these processes. Since my colleagues and I are deeply involved in these topics, it seems like a good opportunity to discuss MIT's Schwarzman College of Computing.[40] We are currently running the "NC Tech Ethics Fellowship" program with the support of NCSOFT, and the enrollment for this course has been oversubscribed. This demonstrates that many students are interested in these issues.

Additionally, the MIT Philosophy Department operates a division called SERC (Social and Ethical Responsibilities of Computing)[41] within the Schwarzman College of Computing, which deals with the social and ethical responsibilities of computing, and three people are currently pursuing postdoctoral studies in this field. As such, there is a significant amount of work being done on technology ethics at MIT, with philosophy playing an essential role. I am confident that similar movements are happening in higher education institutions nationally and globally. The role of philosophy in technology ethics will likely become more prominent in the future. Furthermore, another important indicator is the rapid increase in technology ethics jobs, going from almost zero to a significant number in a short period. In short, it is clear that philosophy encompasses the field of technology ethics.

Yoon: Great. To add another question, do you think the way of introducing technologies that can have a wide-ranging impact on society should change? For example, is an ethical review necessary to check for unforeseen consequences in advance?

Byrne: Yes, of course. We already have ethical review systems in place for all kinds of scientific research, such as research ethics committees. In my opinion, introducing such a system in the technology domain would be a good idea. Incidentally, MIT has posted many case studies on technology ethics on the SERC website.[42] The research results are freely available and continue to be added. This would be a useful resource for those who want to know more about these issues.

Yoon: You're right. It's not only a pleasure to see various case studies continue, but it's also gratifying, and thankfully, we can access such resources anytime and anywhere through internet technology. Thank you. It's also very meaningful to be able to support MIT's efforts in various ways related to the topic of technology ethics. I hope that the field will develop even faster through the exchange between NCSOFT and MIT.

Byrne: Thank you. I sincerely appreciate your support.

Yoon: We are also grateful for the efforts and achievements of you and MIT in this ongoing process. It was more than meaningful today as we were able to review that process through our conversation. Thank you.

ACKNOWLEDGMENTS

I want to thank my guests and colleagues who agreed to be interviewed for this book. Your insights and candor create the backbone of this book. I would also like to thank Nayoung, MJ, and Ron for helping me create the content and bring the story to market. Finally, thank you to my family, friends, and colleagues who have encouraged me to share my thoughts here; your support and encouragement mean the world to me.

ENDNOTES

1   Rob Toews, "8 Leading Women in the Field of AI," Forbes, December 13,2020, https://www.forbes.com/sites/robtoews/2020/12/13/8-leading-women-in-the-field-of-ai/?sh=2963a9565c97.

2   John McCarthy, known as the father of artificial intelligence, coined the term "artificial intelligence" and introduced its foundation in 1956 at an academic conference at Dartmouth.

3   The Turing Award, considered the Nobel Prize of computer science, is awarded annually by the Association for Computing Machinery (ACM) for contributions of lasting and significant technical importance to computer science.

4   A conference organized by ACM where fairness, accountability, and transparency of computer science engineering related research are discussed. The first ACM FAccT outside of North America and Europe was held in South Korea in 2022.

5   Stanford Institute for Human-Centered Artificial Intelligence, "Summary of AI Provisions from the National Defense Authorization Act 2021," https://hai.stanford.edu/policy/policy-resources/summary-ai-provisions-national-defense-authorization-act-2021.

6   US Government Publishing Office Washington, "William M. (Mac) Thornberry National Defense Authorization Act for Fiscal Year 2021," ordered to be printed December 2020, https://docs.house.gov/billsthisweek/20201207/CRPT-116hrpt617.pdf.

7   National Institute of Standards and Technology U.S. Department of Commerce, "AI Risk Management Framework," accessed October 2022, https://www.nist.gov/itl/ai-risk-management-framework.

8   Stanford University Human-Centered Artificial Intelligence, "Summary of AI Provisions from the National Defense Authorization Act 2021," accessed October, 2022, https://hai.stanford.edu/ summary-ai-provisions-national-defense-authorization-act-2021.

9   Congress.gov, "All Information (Except Text) for S.3890—National AI Research Resource Task Force Act of 2020," accessed October 2022, https://www.congress.gov/bill/116th-congress/senate-bill/3890/ all-info.

10   Stanford University Human-Centered Artificial Intelligence, "National AI Research Resource," accessed October 2022, https://hai.stanford. edu/policy/national-research-cloud; Steve Lohr, "Universities and Tech Giants Back National Cloud Computing Project," *New York Times*, June 30, 2020, https://www.nytimes.com/2020/06/30/technology/ national-cloud-computing-project.html.

11   Daniel Zhang, Nestor Maslej, Erik Brynjolfsson, John Etchemendy, et al., "The AI Index 2022 Annual Report," AI Index Steering Committee, Stanford Institute for Human-Centered AI, Stanford University, March 2022.

12   The AI Index by HAI is an annual report that collects data on various aspects of AI, such as research and development, the current state of technology, economic impact, education, policies, and ethics. This report is produced through collaboration with various groups in association with HAI.

13   Stanford University Human-Centered Artificial Intelligence, "2019 Fall Conference: AI Ethics, Policy, and Governance," accessed October 2022, https://hai.stanford. edu/2019-fall-conference-ai-ethics-policy-and-governance.

14   A nonprofit organization established in 1920 to protect the rights and freedom of individuals guaranteed by the US Constitution.

15   A nonprofit organization established in 2016 to counter potential discrimination and irrationality from AI technology. They expose prejudice and bias in AI systems and hold social movements to help the general public to learn about AI's influence.

16   National Institute of Standards and Technology (NIST), "AI Risk Management Framework," https://www.nist.gov/itl/ ai-risk-management-framework.

17   Dual revolution refers to the period between the late eighteenth century and the early nineteenth century during the Industrial Revolution and the French Revolution over three times. The term was first coined by the British historian Eric Hobsbawm. The dual revolution brought on the modern age of capitalism and democracy to the West.

18   Robert Booth, "Facebook Reveals News Feed Experiment to Control Emotions," *The Guardian*, November 30, 2017, https://www.theguardian. com/technology/2014/jun/29/facebook-users-emotions-news-feeds.

19   End-to-end encryption (E2EE) is a technology that prevents third parties and even platform providers from accessing the data transmitted between individuals.

20   Gwihoon Cho, "Establishing International-Level Bioethics: Focusing on the Strengthening of the Role of Institutional Bioethics Committees," Bioethics Forum, South Korea: Korea National Institute for Bioethics Policy, 2013, https://www.nibp.kr/xe/?module=file&act=procFileDo wnload&file_srl=223913&sid=9b633a61829549d88ea95cd5b4700 99e; Songyon Cho, "A Review of Bioethics and the IRB in Social and Behavioral Research," *Korean Association of Childcare & Education* 14, no. 2 (2018): 1–17; Hwang Sangik, "The Past and Present of Bioethics in Korea," *Life, Ethics, and Policy* 1, no. 1 (2017): 31–55. doi: 10.23183/konibp.2017.1.1.002; Act on bioethics and safety, Korean Law Information Center, 2021, https://www.law.go.kr/%EB%B2%9

5%EB%A0%B9/%EC%83%9D%EB%AA%85%EC%9C%A4%E
B%A6%AC%EB%B0%8F%EC%95%88%EC%A0%84%EC%97
%90%EA%B4%80%ED%95%9C%EB%B2%95%EB%A5%A0.

21   Formed in 2020, Facebook's Oversight Board consists of twenty global constitution and human rights experts and law professors from US universities like Stanford and Columbia. The board does not cover all decisions made by Facebook, but they cover politically significant or difficult decisions that have the chance to be reversed, giving them the nickname "Facebook's Supreme Court."

22   Stanford Embedded Ethics, "Embedding Ethics in Computer Science," accessed December 1, 2023, https://embeddedethics.stanford.edu/.

23   Professor Barbara J. Grosz (1948–) is a US computer scientist who contributed greatly to the field of AI by developing the first computer conversation system and is renowned for elevating the status of women in the field of science.

24   Daniel Dennett (1942–) is a US philosopher, author, and cognitive scientist who primarily focuses on mental philosophy, philosophy of science, and biophilosophy.

25   Daniel C. Dennett, *Brainchildren: Essays on Designing Minds* (Cambridge, MA: MIT Press, 1998).

26   Codex is an AI system announced in 2021 by OpenAI, an artificial intelligence company and research institute co-founded by Elon Musk and Sam Altman in 2015.

27   Animal machine theory is a part of Descarte's scientific mechanistic worldview, which interprets all animal movements through mechanical principles. He compared the bodies of animals to automatic machines made by humans, claiming that animals were machines created much more intricately by God. However, in the case of humans, he proposed dualism, stating that while the body works like a machine, the soul exists completely separate from the body.

28 HCI (human-computer interaction) is a field of study that primarily focuses on researching effective methods to improve the interaction between computer systems and human users. Its purpose is to explore and develop the optimal user experience when using computer systems.

29 Lei Guo, Jacob A. Rohde, and H. Denis Wu, "Who Is Responsible for Twitter's Echo Chamber Problem? Evidence from 2016 U.S. Election Networks," *Information, Communication & Society* 23, no. 2 (July 20, 2018): 234–251, doi: 10.1080/1369118x.2018.1499793.

30 "The Facebook Files," Wall Street Journal, October 1, 2021, https://www.wsj.com/articles/the-facebook-files-11631713039.

31 Ibid.

32 Wonjae Woo, *The Age of "Like" Murder: Those Who Stand Against Monster Groups* (South Korea: Yangmoon, 2021); Sung-gyu Lee, "Echo Chamber Effect and Social Engineering," *Journal of Media People* 256: 5–6, https://www.pac.or.kr/kor/pages/?p=204&magazine_new=M02&cate=MA01&nPage=1&idx=1050&m=view&f=&s=; Jinhyung Cho and Kyujung Kim, "A Study on the Improvement of Filter Bubble Phenomenon by Echo Chamber in Social Media," *Korea Journal of Contents* 22, no. 5 (2022): 56–66, doi: 10.5392/JKCA.2022.22.05.056.

33 *WSJ*, "The Facebook Files," October 1, 2021, https://www.wsj.com/articles/the-facebook-files-11631713039.

34 Ibid.

35 Ibid.

36 Judith Jarvis Thomson (1929–2020) was a US philosopher, and the trolley problem, through Thomson, became established as a subfield of ethics, experimental philosophy, and psychology, known as "trolleyology."

37   Georgia Wells, Jeff Horwitz, and Deepa Seetharaman, "Facebook Knows Instagram Is Toxic for Teen Girls, Company Documents Show," *WSJ*, September 14, 2021, https://www.wsj.com/articles/facebook-knows-instagram-is-toxic-for-teen-girls-company-documents-show-11631620739.

38   Ibid.

39   The Gender Shades Project is a research project led by Joy Buolamwini at MIT Media Lab, aiming to uncover biases in technological algorithms. This project revealed that facial recognition programs show differences in classification accuracy depending on gender and race.

40   MIT Schwarzman College of Computing, "MIT Schwarzman College of Computing," n.d., https://computing.mit.edu/.

41   MIT Case Studies in Social and Ethical Responsibilities of Computing, "MIT Case Studies in Social and Ethical Responsibilities of Computing," n.d., https://mit-serc.pubpub.org/.

42   Massachusetts Institute of Technology, "MIT Science and Engineering Review," accessed October 9, 2023, https://mit-serc.pubpub.org.

9 798891 880276